T0173277

NEW
INTEGRATED
SCIENCE
FOR THE CARIBBEAN

A Lower Secondary Course

BOOK 2

John Steward
Steve West
Eugenie Williams

Orders: please contact Hachette UK Distribution, Hely Hutchinson Centre, Milton Road, Didcot, Oxfordshire, OX11 7HH.
Telephone: +44 (0)1235 827827. Email education@hachette.co.uk. Lines are open from 9 a.m. to 5 p.m., Monday to Friday.
You can also order through our website: www.hoddereducation.co.uk

First published 1989 by Pearson Education Limited
Second edition 2000
Reprinted 2017

Published from 2015 by Hodder Education,
An Hachette UK Company
Carmelite House
50 Victoria Embankment
London EC4Y 0DZ
www.hoddereducation.com

Impression number: 10 9 8 7
Year: 2023

ISBN 978-0-582-33263-8
Printed and bound by CPI Group (UK) Ltd, Croydon, CR0 4YY
GCC/09

Acknowledgements

Authors of the first edition: Lucy Durgadeen, Vilma McClenan, Steve West and Eugenie Williams with the editorial advice of John Steward.

The Publishers are grateful to the following for permission to reproduce photographs:
Alcan Jamaica Company for Figure 9.3c; All Sport (Clive Brunskill) for Figure 3.1; Astra Zeneca for Figure 9.10g; Avecia Ltd for Figure 9.10e; British Gas Group for Figure 2.61; Centre for Alternative Technology for Figure 9.1; GeoScience Features Picture Library for Figures 2.32 and 3.25; Greenbrook Electrical PLC for Figure 4.19 (bottom); Griffin and George for Figures 2.12, 2.41 (bottom left), 2.42, 4.8, 6.48b and 7.22; Hutchison Library for Figures 6.41 (Liba Taylor), 7.11b (Michael Kahn), 8.16 (Sarah Errington, top; Godfrey Morrison, bottom) and 8.18 (J. Henderson); London Symphony Orchestra, bottom (Keith Saunders) for Figures 6.43b and 6.51 (top); On the Slide Photographic Library (John Bares) for Figures 1.3, 1.6, 1.7, 2.5, 2.15b, 2.16, 2.18, 2.22, 2.29, 2.34, 2.41 (top and bottom right), 2.52, 2.62, 3.33, 4.1 (right), 4.6, 4.17, 4.18, 4.20b and c, 4.28, 5.10, 5.11, 5.12, 5.30, 6.19, 6,20, 6.42, 6.43c, 6.48a and d, 6.49, 6.51 (bottom), 7.17 (top), 7.35, 7.41, 8.8, 8.12, 8.13, 9.3a, 9.8 and 9.10a, b, d, f; Philips Domestic Appliances and Personal Care for Figures 4.1 (left) and 6.48c; Powell Marketing Group for Figure 9.2; Royal Festival Hall for Figure 6.59; Science Photo Library for Figures 1.1 (Richard T. Nowitz), 1.2 (Will & Deni Mcintyre), 2.7, 2.15c (Steve Horrell), 2.20 (Mark Burnett), 2.21, 2.23 (Sinclair Stammers), 2.24 (Marcelo Brodsky), 3.36 (Ron Church), 4.3 (Keith Kent), 4.7 (Geoff Tompkinson), 4.19 (top) (U.S. Dept of Energy), 4.26, 5.1 (NASA), 6.10 (David Nunuk), 7.20 (Sinclair Stammers), 8.20 (Simon Fraser), 9.3b (Damien Lovegrove), 9.4 (Gary Parker), 9.7 (Rosenfeld Images Ltd), 9.10c (Pascal Goetgheluck) and 9.11 (John Mead); Severn Trent Water for Figure 2.15a; Shell Photographic Library for Figure 8.17; David Simson for Figures 6.1, 6.38, 6.39, 7.1, 7.11a and 8.1; Alex Smailes for Figures 7.42 and 9.5; Sony Consumer Products Group for Figure 6.43a; Trinidad and Tobago Tourism Office for Figure 7.6; J. Tyndale-Biscoe for Figure 8.14.

The cover photograph was taken by Mark Wilson. It features students from Bishop Anstey High School and Queen's Royal College, Trinidad. The students are studying a transect to deduce how the plant species and growth are affected by changes in soil conditions and other factors along the line of the transect.

We regret that we have been unable to trace the copyright holders of Figures 4.20a and 8.19 and would welcome any information that would enable us to do so.

Introduction for the Student

The aim of this book is to help you to learn, understand and appreciate science. The book does this by showing you how science relates to your everyday life, and by showing you that science is fun. Most importantly, it helps you develop the skills you will need to be a good scientist.

The book contains 9 study **units**. Each unit is clearly set out, beginning with a series of brief **objectives** that summarize what you will learn in the unit. These objectives are followed by **sub-units** that deal with a particular theme within the unit. The sub-units often begin with one or two questions. These questions are designed to make you think about why things around us might be the way they are – which is exactly what a scientist does.

Within the sub-units are many **Activities**, which are designed to help you learn science by finding things out for yourself – using the skills of a scientist. You will normally do these acitivities with other classmates, and with the help of your teacher. But there are also **On your own** exercises. These are small activities that you can do in your own time as homework. They will help to build your confidence in your abilities as a scientist.

Diagrams, photos and illustrations will help you to appreciate the science that goes on around you every day. Some of the activities and objects shown in the pictures are things you may have seen somewhere before, or things that you may even know something about already. This book will show you the scientific aspects of those objects and activities.

Each unit ends with a useful **summary** of the main scientific points you will have learnt in the unit. This is followed by a series of **questions** to test your understanding of the unit.

At the back of the book there are **End of Term Tests** that will help you assess how your work is progressing as the year goes on.

Using this book, you will discover that the best way to learn science is to have an inquiring mind about the world around you. That curiosity can then be put to scientific use by exploring and experimenting, to test your scientific ideas. With this book, you will begin learning how to do this – you will start to learn to become a scientist.

Contents

UNIT 1 Scientists and skills

OBJECTIVES

- Identify the skills that scientists use to carry out scientific investigations
- Discuss the importance of experimental design and the role of variables
- List the guidelines to follow when reporting an experiment
- Explain the importance of graphs in presenting and interpreting data, as well as for making predictions in experiments

If several people are given the same information at the same time, they are likely to come to quite different conclusions. How often have you heard people watching a cricket match argue about who the best batsman is? One person might argue that the best batsman is the one who scored the most runs in that particular match. Another person might say that the overall number of runs the batsman scored in the last test series is more important. Yet another might consider how many sixes and fours a batsman scored in a particular time. In other words, different people have different ways of gathering information and of interpreting facts. You have already learnt about the special way in which scientists gather and interpret facts: the scientific method. In this unit we will first investigate in more detail some of the skills scientists need when they use the scientific method; we will then use some of these skills in activities.

Figure 1.1   *A researcher examining test tubes*

1.1 The skills of a scientist

Figure 1.2 *Scientists carry out many types of experiments. They collect data, interpret the data and draw conclusions*

How many skills do you think a scientist needs to use when carrying out an investigation? You will see from the list below that it is a surprisingly varied number.

Observing: using one or more of the senses to find out about objects or events in the environment.

Measuring: using units of measurement and measuring apparatus to compare an unknown quantity with a standard quantity.

Comparing and contrasting: recognizing ways in which objects or events are alike or different.

Classifying: grouping objects or events according to their observed properties.

Recognizing patterns and relationships: identifying ways in which objects or events are related.

Using space and time relationships: recognizing and describing relationships between objects and events in terms of their shape, position, duration and sequence in time.

Hypothesizing: suggesting answers to questions or problems that can then be tested by experiments.

Dr Maura P. Imbert is Programme Leader of the Agro Industrial Products Programme at the Caribbean Industrial Research Institute. For the past ten years she has been doing extensive research on two projects. The first is the development of natural preservatives from local herbs and spices. These products, extracted from plants such as Spanish thyme, can be used as partial or sole replacements for sodium benzoate, which can be harmful to humans if ingested in large quantities.

Dr Imbert has also been working on the production of natural insecticide from two plants, neem and Persian lilac. These plants belong to the mahogany family. They have seed kernels from which has been extracted a very active insecticide. The insecticide has given good results in the laboratory against corn and sugar cane borers. At present, Dr Imbert is conducting field trials on the insecticide and is currently engaged in research on medicinal plants, particularly Aloe vera.

Predicting: using previous knowledge and recognizing possible causes and effects to suggest in advance what the outcome of an event or process will be.

Experimenting: designing and performing a set of test procedures under controlled conditions to obtain reliable information.

Interpreting: explaining the meaning or significance of information collected.

Inferring: arriving at a conclusion based on the observations made during an experiment.

Communicating: giving information through pictures, diagrams graphs, maps and written or oral descriptions. This requires the careful recording of observations and experiments.

Figure 1.3 *Careful presentation of results helps scientists communicate their conclusions*

Figure 1.4 *Dr Dellimore, a biophysicist from St Vincent*

Dr Jeffrey W. Dellimore, from St Vincent and the Grenadines, is a biophysicist. He conducted research into problems of blood flow in human beings, paying special attention to the behaviour of red blood cells under various conditions. In 1979, he received the Canadian Heart Foundation 'Visiting Scientist Award' for his work in this area.

Dr Dellimore is now involved in corporate planning and policy formulation with the Caribbean Development Bank. He was instrumental in the formulation of the 'Caribbean Regional Energy Action Plan' and 'A Policy for Science and Technology in the Caribbean'.

As a child, Jeffrey Dellimore was keen on science and technology. From the age of 12, he was a radio amateur (HAM) operator using radio transmitters that he had designed and built from components salvaged from old radios.

Scientific attitudes

In addition to these skills, there are certain attitudes which scientists show. A good scientist always *shows curiosity* about things in the environment, and therefore wants to try to solve problems. Scientists *respect all living things,* and are willing to use correct and safe procedures during their investigations. They also *respect other people's ideas,* and are willing to discuss and share ideas with others. They should be *open-minded* and want to *separate fact from opinion.* Finally, they are always willing to *persevere;* that is, to continue working on problems even when they are not getting the kinds of results expected.

However, scientists are also human beings and are not perfect! They do not always show all these attitudes but they aim to.

1.2 Experimental design and variables

Have you ever thought, when doing an experiment, just how many different conditions can affect the result?

Conditions which can change in an experiment are called **variables.** When you design an experiment, you must try to change only one variable at a time, and observe the effect that this change has. If you change many variables at one time, you cannot tell which change has caused the effects you observe.

For example, imagine you are investigating the note that is produced when a guitar string is plucked. If you

Figure 1.5 *Variables light, water, plant food and temperature*

changed both the length and the tension (tightness) of the string, you would find that the pitch of the note produced is different. But you could not tell if it was the change of length or the change of tension, or both, which caused the pitch of the note to change. In an experiment it is better to change only one variable (the length, say) and see what effect that has on the note produced. Then, in a separate experiment, you can change the other variable (the tension) while keeping the length the same. In this way, you can see the effect of changing each variable.

The variable that is changed by the experimenter is called the **independent** or **manipulated variable**. In the experiment mentioned above, this would be the length of the guitar string. The variable which changes as a result of this is called the **dependent** or **responding variable**. This would be the pitch of the note. Variables which are kept the same throughout the experiment are called **controlled variables**.

In Book 1, Unit 5, you learnt that yeast breaks down glucose to produce carbon dioxide and ethanol: a process called **fermentation**. When yeast is used in baking, the carbon dioxide causes the dough to rise.

Figure 1.6   *Types of commercially produced yeast*

Figure 1.7   *Dough, containing yeast, before and after baking*

While helping in the kitchen, your friend observed that his mother always mixed the yeast in **warm water** when making bread. Based on that observation over a period of time, your friend made this statement: 'Yeast is more active in warm water than in cold water'. This statement is a hypothesis which you are going to test.

ACTIVITY 1.1   TESTING A HYPOTHESIS

You are going to test the hypothesis that yeast is more active in warm water than in cold water.

You will need some commercial yeast (not instant yeast), sugar, ice-cold water, warm water, two beakers, a measuring cylinder, a ruler, a thermometer and a watch.

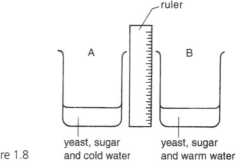

Figure 1.8

A — ruler — B

yeast, sugar and cold water   yeast, sugar and warm water

❶ In beaker A place 10 ml of cold water and one level teaspoon of sugar. Dissolve the sugar in the water and then add 15 ml of cold water. Record the temperature.

❷ In beaker B place 10 ml of warm water and one level teaspoon of sugar. Dissolve the sugar and then add 15 ml of warm water. Record the temperature.

9

3 To each beaker add one level teaspoon of commercial yeast and stir.

4 Place the beakers side by side on the bench with the ruler between them and measure the height of the mixture after five minutes.

The amount of yeast is a variable you controlled.

List three other variables that were controlled in the experiment.

What is the independent variable?

What is the dependent variable?

Notice that the independent and dependent variables are usually included in the hypothesis.

From the activity give examples of the use of the following skills: classifying, measuring and space–time relationships.

State the inference you made based on what you observed about the activity of the yeast.

Did the results of your experiment support your hypothesis?

Imagine you saw two people about to make bread, and one was using cold water and the other warm water. Make a prediction about which one would take longer to rise – the bread made by cold water or the bread made by warm water.

The process you went through in this activity is an example of applying the scientific method. Use of scientific skills and methods enables us to obtain new scientific knowledge in a logical way.

1.3 Reporting on an experiment

In order to make details of their experiments available to others, scientists record their experiments in reports. A report enables other scientists to check whether they think the experiment has been well designed and carried out; to repeat the experiment themselves if they want to; to find out what the results were; and to find out what conclusions have been drawn from the results. Most scientific reports follow the pattern of the following headings:

Aim
Materials and equipment
Method or procedure
Result

Interpretation of result
Conclusion

Aim: This is a general description of what you set out to do or show, e.g. to test the effect of saliva on starch. You should also indicate what your experimental hypothesis is, i.e. what do you think or expect might happen?

What was the aim of your last activity?

Materials and equipment: Here you list all the materials, equipment or apparatus used in the activity. For example, materials used in your activity included sugar and yeast.

Give four other items from your last activity that you would list under this heading.

Method/procedure: This is a concise account of the steps followed to obtain the results. It may include diagrams of the materials and apparatus as set up.

Would it be helpful to include a diagram of the procedure in a report on your last activity?

Result: Here you present the information you collect. This may be presented in words, as a table, a diagram and/or a graph.

A table of your results in the last activity might look like the table below. Notice that the table must have a **number** and a **title**. Units, e.g. cm, must be written in parentheses at the head of the column. The following table shows the rise in height of yeast after five minutes.

Beaker	Height (cm)
A: cold water	1
B: warm water	2.5

Interpretation of result: Here you examine the results to determine any patterns and/or relationships. In your activity, the beaker in which the solution was higher was the one in which the yeast was more active.

Conclusion: Contains suggested explanations of your results reached by considering your results and the variables in the experiment. You should usually indicate whether the hypothesis was supported or not. For example, 'yeast is more active in warm water than in cold water' would be a conclusion to your last experiment.

You may also state any special precautions that were taken during the experiment.

Make a hypothesis about the effect of sugar on the activity of yeast.

Design and carry out a simple experiment to test the hypothesis.

Write a report of your experiment.

1.4 Presenting results

In Book 1 you often used tables to present your results. You will find out shortly that graphs are also very useful for presenting numerical results. Graphs communicate information more effectively than tables, because they make it easy to see how the independent and dependent variables of an experiment are related. The table below shows the temperature of water in a container as the water is heated and the temperature is measured at one-minute intervals.

time (min)	0	1	2	3	4	5	6	7	8	9	10
temperature (°C)	15	26	36	47	57	67	78	88	98	100	100

This information can be shown on a graph with time the independant variable on the horizontal x axis and temperature the dependant variable on the vertical y axis. Each point marked ✕ on the graph shows the temperature of the water and the time when the temperature was measured.

When ✕s have been marked for each measurement, the marks can sometimes easily be joined up with a straight line or a curved line. If it is not possible to see that the marks would clearly form a straight line or a curved line if joined together, it is acceptable to draw a straight line or smooth curve which goes very near the points, but not through all of them. This is described as drawing a line or curve of best fit. It is used because measurements made in experiments are never exactly correct. The graphs in Figure 1.9 show the temperature increase of water over a period of time.

A line of best fit is drawn so that there are about the same number of plotted points on either side of the line or curve.

From the graphs shown here, it is easy to see that the temperature of the water rises until it reaches 100°C, then it stops rising.

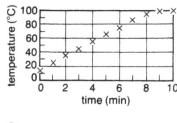

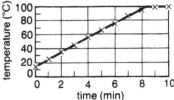

Figure 1.9

ACTIVITY 1.2 PRESENTING RESULTS

You are going to carry out an activity similar to Activity 1.1, but present your results differently.

You will need some commercial yeast, sugar, cold water, warm water, two beakers, a measuring cylinder, a ruler, a thermometer and a watch.

Repeat steps 1 to 3 of Activity 1.1.

Measure the height of the mixture in each beaker after one minute and then each minute for five minutes.

Record your results in a table like this one:

Beaker	Height (cm) after					
	1 min	2 min	3 min	4 min	5 min	6 min
A cold water						
B warm water						

Record your results on two graphs with the axes as shown below.

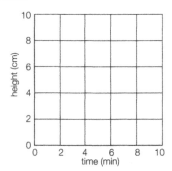

Figure 1.10

5 Name the independent variable. On which axis is it placed? Name the dependent variable. On which axis is it placed?

Force (N)	Length (cm)	Length/force $\frac{y}{x} = m$	$m \times x = y$
0	0		
1	1.5	$\frac{1.5}{1} = 1.5$	$1.5 \times 1 = 1.5$
2	3.0	$\frac{3.0}{2} =$	$1.5 \times 2 =$
3	4.5	$\frac{4.5}{3} =$	$1.5 \times 3 =$
4	6.0	$\frac{6.0}{4} =$	$1.5 \times 4 =$
5	7.5	$\frac{7.5}{5} =$	$1.5 \times 5 =$

1.5 Interpreting graphs

We will now take a closer look at graphs and explore their importance in scientific investigations.

Estimating with graphs

Graphs not only communicate information effectively, they are also used to estimate values or readings that are not obtained experimentally. For example, you can use the graphs in Figure 1.9 on page 11 to estimate the temperature of the water after 2.5 minutes, or 3.5 minutes. You can also predict the temperature after 12 minutes, providing the temperature follows the same trend that it showed during the earlier minutes. Notice that you obtained readings between plotted points and also readings beyond the limits of the plotted points. The plotted points were obtained experimentally, but your readings were not.

Shapes of graphs

Shapes of graphs tell us how the variables on the graph are related in a quantitative way, and serve as a useful means of interpreting data tables. Mathematical equations can also be obtained from these relationships, and values for a wide range of variables can be obtained without performing the experiment to get the data.

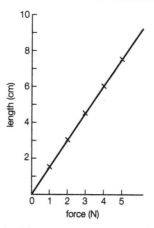

Figure 1.11   *Graph of force against spring length*

  STRAIGHT LINE GRAPHS PASSING THROUGH THE ORIGIN

The table given on the right and the graph in Figure 1.11 show the effect of increasing force on the length of a spring.

Copy and complete the table in your note books. Observe from the table that:

1 When the force is doubled, the length is also doubled. Both variables increase by the same factor. The variables increase **proportionately**. Look at the

next column in your table, where length is divided by force. The result is the same for all the values. Scientists and mathematicians use the letter '*m*' to show that this result, called a **quotient**, is constant. For *any* point on the graph, dividing *y* by *x* always gives the same number, 1.5, which we call '*m*'. This tells us that the whole graph can be represented by one mathematical **equation**, $y/x = m$. (The word 'equation' comes from the word 'equals'.)

2 From this equation comes another one, $y = mx$. The graph that represents this mathematical relationship between the *x*-variable and *y*-variable, and that passes through the origin, is called a straight-line graph. '*m*' is the value of the slope of this straight line.

3 The relationship between the two variables of this graph is called a **direct relationship** because both variables undergo the same change proportionately.

Straight-line graphs not passing through the origin

Some straight-line graphs have a line that does not pass through the origin, but crosses the y-axis instead. The general equation for this type of graph is

$$y = mx + c$$

where:

m = slope of the line

c = the point where the line crosses or intercepts the y-axis.

The general equation can be used to calculate any value of x or y, as long as the values of m, c and either x or y are known.

Example

I would like to find the mass (y) that would be added to a rat's weight if it consumed 2000 calories (x).

I know that $y = mx + c$ for these types of graphs so I can calculate that $m = (y - c)/x$.

I can then work out the value of m using any x and y co-ordinates (see graph) that appear on the line of the graph.

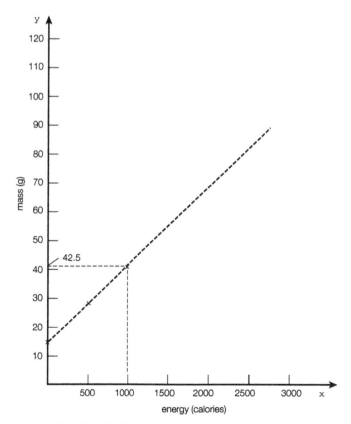

Figure 1.12 *Graph of the mass gained by a rat against calories consumed*

Therefore, $m = (42.5 - 15)/1000 = 0.0275$

Therefore, $y = 0.275 \times 2000 + 15$

$= 55 + 15$

$= 70$

So 2000 calories of food will cause the rat to gain 70 g in mass.

The graph of a curve

Another type of relationship between variables is one where one variable *decreases* and the other *increases* by the same factor. This type of relationship is called an *inverse relationship*. The graph of this relationship is a curve that is called a hyperbola. The table and graph below show the effect of increased pressure upon gas volume. Both table and graph show an inverse relationship.

x Pressure (atm)	y Volume (ml)	$x \times y$ $(P \times V)$
0.10	800	80.0
0.20	400	80.0
0.40	200	80.0
0.60	133	78.8
0.70	114	79.8
0.80	100	80.0

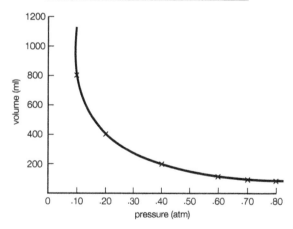

Figure 1.13 *Effect of increased pressure on the volume of a gas*

Note that $P \times V$ always gives the same value, or constant. So the mathematical formula obtained from this relationship is:

$x \times y = k$ or $y = k/x$, where k represents the constant (just like m represented the constant earlier).

Here are some guidelines to follow when constructing your own graphs:

1 Identify the independent and dependent variables.

Place the name of the independent variable on the *x*-axis and the name of the dependent variable on the *y*-axis. Name the axes. Place the unit of measurement for each variable in parentheses.

2 A title that clearly includes the independent and dependent variables should be written at the top of the graph.

3 Determine the interval scale for each axis by first subtracting the smallest value from the largest one. Then divide the difference by the number of intervals you want.

4 Select numbers for intervals that are easy to subdivide, for example multiples of two or five.

5 Use a pencil with a sharp point for plotting data points. Draw the best line or curve of fit with approximately the same number of points above the line as below.

Summary

These are some of the things you have learnt in this unit:

● Scientists use many skills in their work and show certain attitudes.

● Experiments must be carefully designed so that the scientist can identify the cause of his or her observations.

● Scientists must communicate with others about their work if it is to be useful.

● A scientific report includes facts and opinions, and the two must be clearly separated.

● Results can be shown in words, graphs, diagrams or tables to make them clear to a reader.

● Graphs are very useful for presenting and interpreting data, as well as for making predictions about how one variable has affected another, even if you didn't record that particular result when doing your experiment.

QUESTIONS

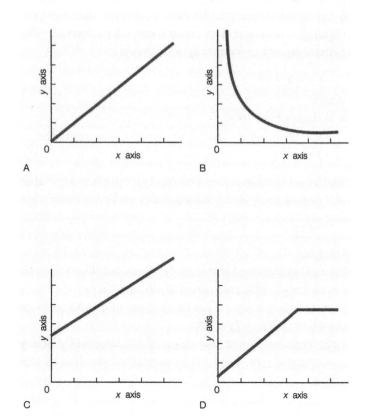

Figure 1.14

Questions 1–6 are about the graphs shown above:

1 In which graph does the value of *y* increase, then remain constant?

2 Which graph represents the relationship $y = mx + c$?

3 In which graph does *x* increase as *y* decreases?

4 In which graph does *y* increase as *x* increases?

5 In which graph does $y = k/x$?

6 In which graph does $y = mx$?

7 The following hypotheses are to be tested by experiments:
'Iron is a better conductor of heat than copper.'
'Iron reacts more vigorously with acids than copper does.'
'A dull black container absorbs more heat than a shiny light container.'
'Woodlice always move away from the light.'
'Bean plants grow towards the light.'

In each case

(i) Describe the experiment you would carry out.
(ii) What is the independent variable?
(iii) What is the dependent variable?
(iv) List the variables you would control.

8 The graph below shows the volume of hydrogen given off when zinc and hydrochloric acid react together.

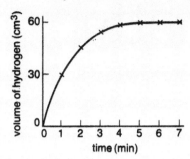

Figure 1.15   *The release of hydrogen over time*

(i) How much hydrogen has been given off after three minutes?
(ii) Name the independent variable and dependent variable.

9 The graph below shows the heights of two girls as they grow up.
(i) Which girl is taller when they are ten years old?
(ii) At what age are the girls the same height?

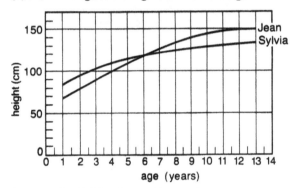

Figure 1.16   *Height of two girls over time*

10 The graph below shows the height of water in bottles of varying shapes and volumes. Use the graph to answer the following questions:

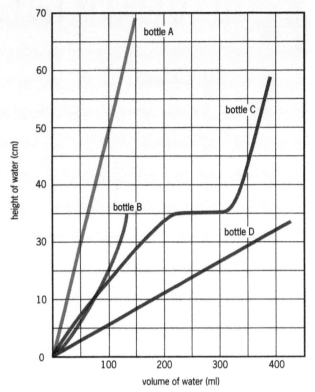

Figure 1.17   *Height of water in bottles of varying shapes and volumes*

(i) Which bottle has the largest volume?
(ii) Which bottle(s) most likely have straight sides?
(iii) Which bottle is the shortest?
(iv) Which bottle has an irregular shape?

Investigating matter

OBJECTIVES

- Learn about the particulate nature of matter
- Define atoms, which are the building blocks of the many groups of matter
- Identify elements, their classification and their patterns of behaviour
- Learn about the structure of atoms
- Learn about solutions and their various properties
- Learn about acids and bases, their uses and their patterns of behaviour

What is matter made of? What are the different ways in which matter can be grouped? These are just two of the questions we can ask when we look around us, and realize that we are living matter and that we are surrounded by other living and non-living matter. In this unit we will discover that scientists have many ways of describing and grouping matter. Matter is more than just solids, liquids and gases – it is made up of even smaller parts called atoms and molecules.

How many forms of living and non-living matter do you see in this photograph?

Figure 2.1 *Living and non-living things are shown here*

2.1 What is matter made of?

Scientists have offered answers to the above question after observing the behaviour of several substances in the solid, liquid and gaseous states. To help us understand the answers scientists have offered, let us now carry out two activities which use two different solids – cane sugar and smelling salts.

 ACTIVITY 2.1 INVESTIGATING THE BEHAVIOUR OF CANE SUGAR IN WATER

You are going to investigate what happens when cane sugar is put in water.

You will need a teaspoonful of cane sugar, a mortar and pestle, some water, a small beaker or glass jar and a glass rod.

❶ Examine a sugar crystal. Describe the sugar crystal in your book.

❷ Crush the crystals into a fine powder. What did the crushing do to each crystal?

❸ Pour the powdered sugar into the beaker containing water, and stir.

❹ Look at the water. Can you still see the powdered sugar?

❺ Place one drop of the liquid on your tongue. What does it taste like?

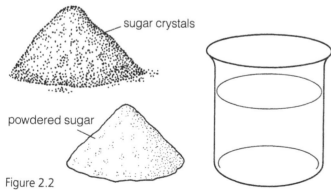

sugar crystals

powdered sugar

Figure 2.2

WARNING: Never taste any substance you use in science lessons unless your teacher says you may do so.

Explaining what happened to the sugar

Your sense of taste tells you that the powdered sugar is still in the water, although you cannot see it. Scientists explain this observation by saying that each small part of the powdered sugar is made up of very, very tiny parts or **particles**. The water caused each part of the powdered sugar to separate into these tiny **particles**. The particles spread throughout the water. The particles are so small that we cannot see them even with a very powerful microscope. However, we can taste them.

 ACTIVITY 2.2 THE EFFECT OF SMELLING SALTS ON LITMUS PAPER

Figure 2.3

You are going to investigate the effect of smelling salts on damp red litmus paper.

You will need smelling salts in a bottle, two pieces of red litmus paper and water.

❶ Dampen the litmus paper with water.

❷ Touch the smelling salts with one piece of damp red litmus paper. Observe and record what you observe.

❸ Hold the other piece of damp red litmus paper about 3 cm above the open mouth of the bottle containing smelling salts for about two minutes. Record what you observe.

Is the colour change the same as when the smelling salts were touched with the first piece of litmus paper? Can you explain this observation?

To help you explain, first answer these questions: Is any matter coming out of the bottle? What is coming out?

Again, scientists use the idea of particles to explain these observations. The solid smelling salts are made up of tiny particles. When the bottle lid is removed, some of these particles escape into the air and reach the red litmus paper. These particles cause the litmus paper to change from red to blue. The particles of smelling salts that escape into the air are in the gaseous state.

On your own

Identify a liquid from which particles escape into the air and carry out a simple investigation to demonstrate this. You may use other students in the class to detect the presence of the particles. The sense of smell may be useful in this activity.

Suggest ways in which you can:

a   make detection of the particles in the air easier
b   increase the number of particles in the air.

> The tongues of snakes and some lizards are very well developed as sense organs. These animals use their tongues to detect, or smell, particles given off into the air by the animals on which they feed.

2.2 The particle idea and states of matter

Matter can be solid such as rock or sugar; liquid such as water or oil; or gas such as air or steam. We call these forms the three states of matter. We now know that solids, liquids and gases are all made up of very tiny particles. What causes a substance to be a solid, a liquid or a gas? The answer is – the way in which the particles of that substance are arranged.

Solids

In the solid state, a substance has a fixed shape, and the amount of space it occupies (its volume) does not change at any one temperature. Solids also have a definite surface which you can see and touch. Solids can be hard or soft, light or heavy. Name some solids that are hard, soft, light or heavy.

The particles in a solid are packed very closely together and are usually in a regular arrangement of rows and columns. Since the particles are close together, they attract each other strongly and tend to be fixed in certain positions. They are not free to move much. They do vibrate, but remain in the same place, like a person standing in one place but shifting weight from one foot to the other.

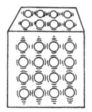

The particles in a solid vibrate about a fixed position.

The particles in a liquid move in all directions beneath the surface.

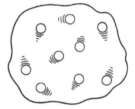

Gas particles move quickly in all directions; there are large spaces between the particles.

Figure 2.4   *Particles vibrating in three states: solid, liquid and gas*

Liquids

A liquid does not have a fixed shape, but has a fixed volume. A liquid also has a definite surface.

In the liquid state, the particles are close together. They attract each other, but not strongly. They slide freely over and past one another. But in a liquid, each particle always remains close to other particles. The free movement of particles in liquids explains why liquids take the shape of their containers, and why they flow, but their volume does not change when the temperature remains the same.

Gases

A gas has no fixed shape or volume. A gas has no definite surfaces – you cannot see the edges of a gas or feel where a gas starts and stops as you can with liquids and solids.

The particles in a gas are far apart. The force of attraction between gas particles is very weak. These particles spread out rapidly and move away from each other. This explains why gases fill any available space. It is also difficult to see most gases.

What is steam?

Steam, or water vapour, is a gas. Steam is invisible. When water boils it changes to steam. The 'steam' we see coming from boiling water is not pure steam, but a mixture of tiny water drops and water vapour.

On your own

1 Copy into your note book and fill in the blank spaces:
   a The name of solid water is _____ .
   b The name of liquid water is _____ .
   c The name of gaseous water is _____ .

2 Describe how particles are arranged in:
   a Solid water.
   b Liquid water.
   c Gaseous water.

3 Explain what happens when water is placed in the freezer section of a refrigerator. (You learned a little about this in Book 1, Unit 7.4, but you should be able to give a more detailed answer now.)

4

Figure 2.5

Place a drop of ink or dye in a glass containing water, as shown in A. Observe the water after 15 minutes. The colour has spread throughout the water as shown in B. What is the ink or dye made up of? What do you think has happened to the particles of ink or dye after 15 minutes?

5 Ask a group of your friends to face the front of the room. Let one person go to the back of the room and open a bottle of perfume. Ask all the others to raise their hands when they smell the perfume.
   Now answer these questions:
   a Did anyone see any perfume as the smell reached his or her nose?
   b Why were you able to smell the perfume?
   c What evidence about particles does this activity provide?

6 a Explain how you can smell the green grass cut by the mower some distance away.
   b List at least five odours you smell every day. What is the source of each odour?

 # Atoms and elements

About 2500 years ago, the Greek philosopher Democritus hypothesized that matter is made up of particles. He called these particles **atoms**. The word 'atom' comes from the Greek word meaning 'uncuttable'. He chose this name because he believed that if he divided a substance in half to get a smaller part, then divided that smaller part in half again and again … he would eventually come to a point when he had a tiny part which could not be cut any further. Democritus said all substances are made up of millions of these 'uncuttable' particles, or 'atoms'.

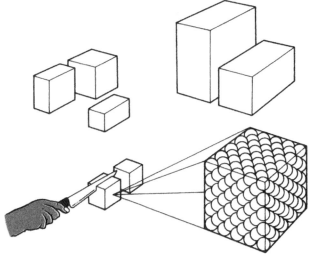
Figure 2.6 *Cutting blocks to show 'uncuttable' particles or atoms*

Figure 2.7 *John Dalton (1766–1844)*

Later John Dalton, an English scientist, suggested that these atoms are not all alike. They differ in size, mass and chemical behaviour. For example, there are atoms of hydrogen and atoms of oxygen; oxygen atoms are bigger and have greater mass than hydrogen atoms.

Dalton said some substances are made up of only one kind of atom. Such substances he called **elements**. Hydrogen, lead, copper, carbon, oxygen and gold are some examples of elements. Hydrogen consists of hydrogen atoms only and oxygen consists of oxygen atoms only. **An element is a substance made up of one kind of atom, and it cannot be broken down into any simpler substance.**

As we have seen, elements contain only one kind of atom. Other substances contain two or more different elements joined together. Air, sea, soil, rocks and all living things are all formed as a result of elements combining in various ways. For example, sugar contains atoms of hydrogen, oxygen and carbon. Sugar is *not* an element because it contains more than one kind of atom. Do you know what elements combine to form meat, water and salt? Look at the picture to find out.

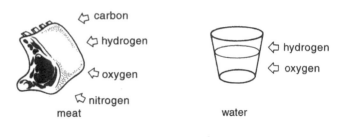
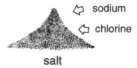

Figure 2.8 *Meat, water and salt contain more than one element*

When two or more atoms join together, they form another type of particle called a **molecule**. For example, two atoms of hydrogen combine with one atom of oxygen to form water. This group of two hydrogen atoms and one oxygen atom is called a water molecule.

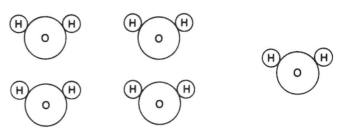

Figure 2.9 *Five water molecules*

Molecules of different compounds have different combinations of atoms in them. Two are shown below.

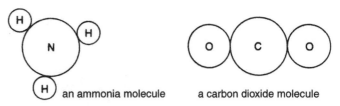

an ammonia molecule   a carbon dioxide molecule

Figure 2.10

Some molecules can be made up of atoms of the same type. For example, oxygen gas (an element) consists of oxygen molecules. Each oxygen molecule contains two oxygen atoms.

an oxygen molecule   a hydrogen molecule

Figure 2.11

A glucose molecule is much larger than the molecules shown above. It is made up of 24 atoms. Some

Figure 2.12 *Models of a glucose molecule (top) and polymers (bottom).*

molecules in animals and plants have thousands of atoms joined up to form them.

2.4 Grouping elements

Of the 106 known elements, 92 occur naturally. Most elements exist in the solid state. Some examples are carbon, calcium, silicon, iron, lead, magnesium and sulphur. About 77 of the 92 naturally occurring elements are solids at room temperature.

A small number of elements are gases at normal room temperature. They are hydrogen, oxygen, nitrogen, chlorine, fluorine, helium, neon, argon, krypton, xenon and radon.

Bromine and mercury are the only two known elements that exist in the liquid state at normal room temperatures.

The ancient Greeks believed that the whole world was made up of four elements: fire, earth, water and air. They thought that the heavenly bodies (stars and planets) were made up of a fifth element: ether.

ACTIVITY 2.3 OBSERVING SOME ELEMENTS

You are going to observe some elements and record your observations.

You will need a few elements, for example, carbon, bromine, copper, iodine, iron, zinc, lead, calcium and sulphur.

WARNING: Do not open the bottle of bromine.

❶ In your book, draw a table like the one shown below and fill it in. One example is done for you.

Name of element	Physical state at room temperature	Appearance
carbon	solid	black shiny surfaces feels light in weight
bromine		

❷ Answer these questions:

a   In which physical state are the elements?

b   Which elements have a similar appearance?

An unusual element

Observe a mercury thermometer. Mercury is another example of an element. What does the mercury in the thermometer look like? How is it different from the other elements you have observed? Mercury is a very poisonous substance so mercury thermometers must always be handled with great care.

Man-made elements
Some elements have been made in science laboratories. These are called 'man-made' or 'artificial' elements. Some have been named after places or famous scientists. Here are the names of a few man-made elements. Whom or what do you think they were named after?
plutonium
berkelium
curium
einsteinium
californium
americium

Metals and non-metals

ACTIVITY **2.4** **CLASSIFYING SOME ELEMENTS**

You are going to observe some elements and classify them to show similarities.

You will need iodine, aluminium, copper sheet or wire, a silver ring, carbon (e.g. coal), sulphur.

1 Examine each element and answer the following questions about each one:

Is it a solid, liquid or gas?
Does it look shiny?
Can you bend it by hammering it?
When hit, does it make a ringing sound?
Can it be drawn out into a thin wire?
Is it brittle?

2 Group the elements into two groups to show their similarities.

3 Compare your groups with those of your classmates.

Metals

Scientists have observed that many elements have similar properties. Some elements were found to have a shiny look, or **lustre**, and to allow heat and electricity to pass through them easily. Many can be easily hammered into thin sheets (they are **malleable**) or made into wires. Elements with these properties are called **metals**. Examples of some metals are gold, silver and aluminium. Mercury is a liquid metal at room temperature.

The most malleable of metals is gold. It can be hammered into sheets so thin that light can pass through them.

Non-metals

There are some solid elements that are dull in appearance and do not allow heat or electricity to pass through them. These elements are also brittle (they break easily) so they cannot be hammered into thin sheets or drawn

into wires. These elements are called **non-metals**. Non-metals can be solid, liquid or gas. Some are listed in the table below.

Solid	Liquid	Gas	
carbon	bromine	hydrogen	neon
silicon		nitrogen	argon
phosphorus		oxygen	krypton
sulphur		fluorine	xenon
iodine		chlorine	radon
		helium	

Now study the table on page 23 that compares the properties of metals and non-metals.

Silver and gold were discovered a long time ago because they usually occur uncombined with other elements in nature. They are considered to be valuable metals and are widely used for jewellery. Copper is a very good conductor of heat and electricity, so copper wires are used in electrical circuits.

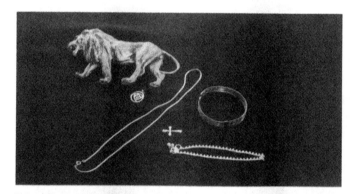

Figure 2.13 *Silver, gold and copper items*

Iodine is a solid at room temperature, not a liquid. The 'liquid' iodine we place on cuts and bruises is solid iodine dissolved in alcohol.

The metals lithium, sodium and potassium are unusual metals. They are so soft that they can be cut with a knife. They tarnish (discolour) very easily and combine readily with oxygen. Sodium and potassium burst into flame when exposed to air. They must be stored in oil to keep them safe.

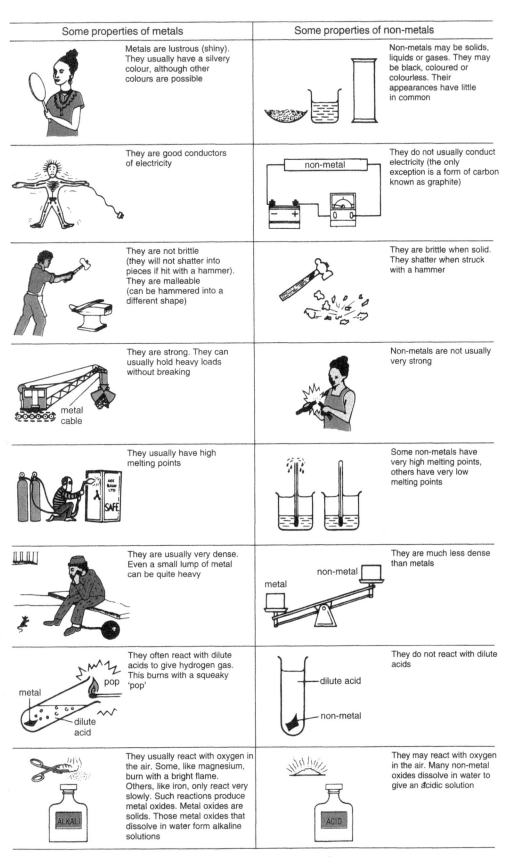

Some properties of metals	Some properties of non-metals
Metals are lustrous (shiny). They usually have a silvery colour, although other colours are possible	Non-metals may be solids, liquids or gases. They may be black, coloured or colourless. Their appearances have little in common
They are good conductors of electricity	They do not usually conduct electricity (the only exception is a form of carbon known as graphite)
They are not brittle (they will not shatter into pieces if hit with a hammer). They are malleable (can be hammered into a different shape)	They are brittle when solid. They shatter when struck with a hammer
They are strong. They can usually hold heavy loads without breaking	Non-metals are not usually very strong
They usually have high melting points	Some non-metals have very high melting points, others have very low melting points
They are usually very dense. Even a small lump of metal can be quite heavy	They are much less dense than metals
They often react with dilute acids to give hydrogen gas. This burns with a squeaky 'pop'	They do not react with dilute acids
They usually react with oxygen in the air. Some, like magnesium, burn with a bright flame. Others, like iron, only react very slowly. Such reactions produce metal oxides. Metal oxides are solids. Those metal oxides that dissolve in water form alkaline solutions	They may react with oxygen in the air. Many non-metal oxides dissolve in water to give an acidic solution

Figure 2.14   *Comparing the properties of metals and non-metals*

1. List four properties of metals. For each property, give an example of how humans have made use of that property in some way. For example, we use the malleability of gold to make jewellery of different shapes.

2. List four household appliances made mainly of metals.

3. Get pieces of four metals such as magnesium, zinc, copper and iron.
   Describe their appearance. Are they hard or soft? Do they float in water or sink in water?
   WARNING: Do *not* try this activity with sodium, potassium or lithium.

4. Some uses of non-metals are shown in the pictures below. Can you think of how we use any other non-metals?

Figure 2.15a *An ozone plant. An ozone molecule is made of three oxygen atoms*

Figure 2.15b *Compounds of fluorine and chlorine are used in toothpastes and bleaches*

Figure 2.15c *A lamp containing iodine gas*

The metal containers used for storing processed foods such as milk, sardines, sausages, etc. are steel cans coated with tin. The coating of tin protects the steel and prevents it from being corroded (eaten away) by the oxygen in the air or by substances in the food.

2.6 Looking at elements

Earlier in this unit you learnt that: (1) an element is made up of one type of atom and (2) elements come together in many ways to form different substances.

In Unit 5 of Book 1 you studied two elements: oxygen and nitrogen. These two gases make up about 98 per cent of the air, which is a mixture of gases. Even though we are surrounded by these gases, we cannot see or smell them because they have no colour or smell. Oxygen also makes up 49 per cent of the Earth's crust, but it occurs combined with other elements.

sandstone
shell
chalk
sand

Figure 2.16 *Many substances in the Earth's crust contain oxygen*

Elements such as gold, silver and copper may be found alone in the Earth's crust, but quite often they, too, are combined with other elements. No matter how elements occur in the environment, we can still place the 106 known elements into two large groups: metals and non-metals.

In this unit, therefore, we will take a close look at two non-metals and a metal. One of the non-metals – hydrogen – is a gas. The other non-metal – carbon – is a solid. The metal – aluminium – is a solid.

Hydrogen: the lightest element

Hydrogen, a gas, is the lightest of all the known elements. The symbol for hydrogen atom is H. A hydrogen molecule is made up of two hydrogen atoms. The symbol for a hydrogen molecule is H_2.

Hydrogen combines with other elements to form a large number of compounds. It is one of the elements which make up the compound water and is a part of many compounds which make up living matter. All acids contain the element hydrogen.

About 80 per cent of the mixture of hot gases which makes up the sun is hydrogen.

ACTIVITY 2.5 PREPARATION AND PROPERTIES OF HYDROGEN

You are going to prepare some hydrogen gas in the laboratory and investigate its properties.

You will need a flask with stopper, a thistle funnel, a delivery tube, test tubes, a trough, dilute hydrochloric acid, water, zinc granules.

1. Set up the apparatus as shown in Figure 2.17. Put a few granules of zinc in the flask before fitting the stopper.

2. Pour dilute hydrochloric acid through the thistle funnel.

3. Collect the bubbles of hydrogen gas over water in test tubes. Put a stopper in each test tube as soon as it is removed. Hold the test tube upside down until the stopper is fitted. Why do you think you need to do this?

4. Collect two or three test tubes of the gas.

5. Put a burning splint into one of the test tubes of hydrogen. What do you observe? Can you see anything on the inside of the test tube?

6. Examine one of the test tubes containing the gas. Note the colour. Take out the stopper. Can you detect any odour?

7. Record your observations by writing down the answers to these questions.

What was the colour of the hydrogen?

Did you detect any distinct odour when the test tube was opened?

Is the gas soluble in water? Give a reason for your answer.

What happened when the lighted splint was put into the test tube with the gas?

How was the hydrogen prevented from escaping into the air through the thistle funnel tube?

Through which piece of apparatus did the gas travel from the flask to the test tube?

Why was the end of the delivery tube above the surface of the acid?

Which was the source of the hydrogen – the zinc or the hydrochloric acid?

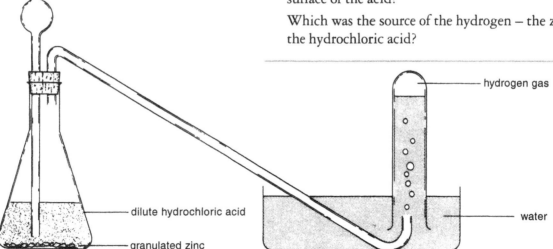

Figure 2.17 — dilute hydrochloric acid — granulated zinc — hydrogen gas — water

You may have noticed some droplets of water on the inside of the test tube in step 5. The compound water is formed when hydrogen combines with oxygen in the air. A mixture of hydrogen and oxygen in the right proportion is very explosive. In 1783 the French scientist Antoine Lavoisier named the gas 'hydrogen', which is Greek for 'water maker'.

On your own

Look back at the guidelines for reporting on page 10, and write up the activity you have just carried out in your notebook.

Hydrogen in margarine

Ask your teacher to take you on a tour of a margarine factory if there is one in your area. Find answers to the questions below. If you cannot visit a factory, try to find the answers from books or by writing to a margarine factory.

a What vegetable oils are used for making margarine?

b Why is hydrogen used in making margarine?

Figure 2.18 *Margarine*

c Where do margarine factories get hydrogen from?

d How do factories prevent hydrogen from exploding?

e What are the main stages in the making of margarine?

The margarine that we use on our bread and for making cakes is made by combining hydrogen with vegetable oils such as sunflower oil, corn oil or peanut oil. The hydrogen causes these oils to change into solid margarine. The process is called **hydrogenation**.

Hydrogen in ammonia

Large quantities of hydrogen are also used to make the gas ammonia. An ammonia molecule contains one nitrogen atom and three hydrogen atoms. Ammonia gas has a very distinctive smell and makes your eyes watery. Ammonia is used to make fertilizers, nitric acid and household cleaners.

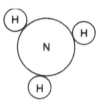

Figure 2.19 *An ammonia molecule*

Hydrogen in balloons and airships

The tyre company *Goodyear* used to make enormous airships which were about 250 m long. Some could carry five small aeroplanes which could take off and even land while the airship was in flight.

Hydrogen was one of the first gases to be used in large weather balloons. Because hydrogen is so light, the balloons lifted quickly into the air. But hydrogen is very explosive when mixed with oxygen. A serious disaster took place in 1937 when the Hindenburg, a giant airship filled with hydrogen, exploded. The explosion took place in Lakehurst, New Jersey, USA, and 36 people were killed.

Helium now replaces hydrogen as the gas used in weather balloons. It is an inert gas, i.e. it does not burn or explode. It is slightly heavier than hydrogen, but still much lighter than air.

Figure 2.20   *A helium balloon*

Henry Cavendish (1731–1810)

Figure 2.21

Henry Cavendish, an English scientist, discovered hydrogen in 1766. He studied the bubbles of gas given off when he dropped a piece of iron into a container of acid. He carefully collected the gas and found that when it burnt, water was formed.

Carbon

Carbon is a solid non-metallic element. It exists uncombined in the earth's crust as **diamond** and **graphite**. It combines with oxygen and occurs as carbon dioxide in the air. It is found in rocks such as limestone, chalk and marble. Fuels such as coal, natural gas, wood, alcohol and gasoline all contain the element carbon.

Carbon is also an important element contained in all living things. It is present in foods that contain fats, proteins and carbohydrates and is thus an important part of protoplasm – the living material of all cells.

Forms of carbon

Carbon is found in nature in pure forms and impure forms. Graphite and diamond are pure forms of carbon: they both contain only carbon atoms, but with the atoms arranged in different ways. Coal, which occurs naturally, and charcoal, which is made by people from wood, are some impure forms of carbon.

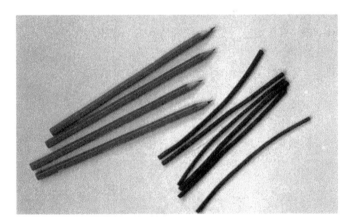

Figure 2.22   *Graphite and charcoal are forms of carbon*

Graphite

Graphite is black, soft and slippery. The substance we call lead in pencils is, in fact, graphite mixed with clay; it contains no lead at all. The name 'graphite' comes from a Greek word meaning 'to write'. To make pencil leads harder, more clay is added. You will investigate some of the properties of graphite later in this unit.

Diamond

Diamond is transparent and glass-like. It is the hardest naturally occurring substance. Diamonds for jewellery are cut (using other diamonds as the cutters) so that

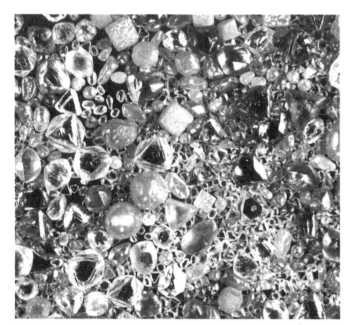

Figure 2.23 *A mixture of natural and synthetic industrial diamonds*

they reflect as much light as possible. Diamonds are also used in industry because of their hardness. They are used in drill bits and saw blades to cut through metals, concrete and granite.

Some styluses used for playing gramophone records are made of diamond. Diamond becomes blunted much less quickly than other materials, so diamond styluses wear records down less than other kinds of styluses.

On your own

Obtain samples of graphite and charcoal and observe their colour, hardness and texture. Copy the table below into your notebook. It describes the properties of various forms of carbon. Fill in the table using your observations and what you know or can find out about diamonds.

Property	Charcoal	Graphite	Diamond
appearance texture hardness			

Coal, charcoal and coke

Coal is an impure form of carbon. It is formed from trees buried deep in the Earth's crust millions of years ago. Coal is dug out of coal mines, and used as a fuel.

Charcoal, another impure form of carbon, is made by heating wood in the absence of air. It is used for cooking, in black polishes and in printers' ink.

Coke, a hard, grey solid, is made by heating coal in the absence of air. It is used as a fuel.

Figure 2.24 *Smoking oven during the production of charcoal in Brazil*

ACTIVITY 2.6 PHYSICAL PROPERTIES OF GRAPHITE

You are going to investigate some of the physical properties of graphite.

You will need a pencil with a sharp point and a piece of white paper.

❶ Rub your pencil up and down on the piece of paper until you make a darkened patch about 2 cm in diameter.

❷ Rub your finger over this darkened patch, and describe how the patch appears and feels.
   Why is graphite useful to lubricate things? (The word 'lubricate' means to make slippery. For example, we lubricate our bicycles when we use oil to make them move more smoothly.)

❸ Examine the point of your pencil.
   Has it become shorter or blunter?
   What does this observation indicate about:
   a   the hardness of graphite, and
   b   how graphite is made up?

ACTIVITY 2.7 THE EFFECT OF HEAT ON GRAPHITE

You are going to investigate what effect heat has on graphite.

You will need a graphite rod, a pair of tongs and a Bunsen burner.

❶ Use a pair of tongs to hold a graphite rod in a Bunsen flame for two minutes.

❷ Allow the rod to cool, then examine it to see if any changes have taken place. Is graphite affected by heat?

Graphite as a lubricant

There are advantages in using graphite to lubricate moving parts of machinery.

Oil or any greasy substance such as *Vaseline* can be used to grease moving parts of machinery or metal objects. But if you examine the hinges of a car door or the chain of a bicycle, which are lubricated with oil, you will notice how much dust sticks onto them. Oil and greasy solids also burn when moving machine parts become very hot.

To solve the problem of dust and burning oil, machine parts are lubricated with graphite. Graphite does not attract dust and does not burn if the moving parts that it lubricates get very hot.

ACTIVITY 2.8 DOES GRAPHITE CONDUCT ELECTRICITY?

You are going to investigate whether graphite will conduct electricity.

You will need a graphite rod, two electric cells, connecting wires, crocodile clips, a bulb, a switch and a length of copper wire.

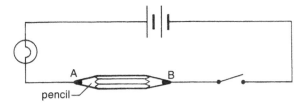

Figure 2.25

❶ Set up the circuit shown in Figure 2.25.

❷ Predict what will happen if an electrical conductor is placed between A and B and the switch is closed.

❸ Place some copper wire (an electrical conductor) between A and B and close the switch. Was your prediction correct?

❹ Graphite is a non-metal. Predict what will happen if graphite is placed between A and B and the switch is closed.

❺ Place a graphite rod between A and B and close the switch.
What do you observe?
Was your prediction correct?
How do you interpret your observations?
What conclusions do you come to?

From what you have learnt about metals, a reasonable hypothesis would be that metallic elements conduct electricity and non-metallic elements do not. This hypothesis would lead you to predict that the bulb would not light in step 5 above. However, you have observed that the bulb *did* light. You must therefore modify your hypothesis.

A new hypothesis could be 'Metallic elements conduct electricity and most non-metallic elements do not. However graphite is an exception as it is a non-metallic element which does conduct electricity.' This observation leads to further questions to try to find a more general rule which would explain why graphite is an exception.

ACTIVITY 2.9 THE EFFECT OF CHEMICALS ON GRAPHITE

You are going to investigate the effect of acid, bleach and alkali on graphite.

You will need three test tubes, a test tube rack, three graphite rods, dilute hydrochloric acid, bleach (Chlorox) and dilute sodium hydroxide solution.

WARNING: Be careful when handling these liquids.

❶ Half fill one test tube with hydrochloric acid; half fill the second test tube with bleach; and half fill the third test tube with sodium hydroxide solution.

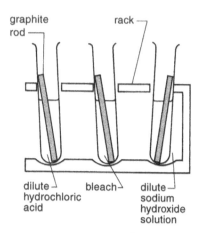

graphite rod

rack

dilute hydrochloric acid

bleach

dilute sodium hydroxide solution

Figure 2.26

2 Place a graphite rod in each test tube and leave them for 30 minutes.

3 Carefully remove each rod, wash it with tap water and examine it to see if any changes have taken place.

Is graphite affected by these chemicals?

Aluminium

Aluminium is the most abundant metal element on earth, and makes up about 8 per cent of the Earth's crust. It is found in rocks and clays, and occurs combined with oxygen and silicon. The main aluminium ore is bauxite: a reddish, clayey-looking material. Bauxite is mined in five Caribbean countries: the Dominican Republic, Guyana, Haiti, Jamaica and Suriname.

Aluminium is a light, very shiny metal. The pure metal is not strong. However, when it is melted and mixed with other metals such as **copper** and magnesium, it forms alloys which are very strong and light.

Aluminium is a reactive metal, but its surface does not wear away or **corrode** by chemical action with water or air. This is because it reacts with oxygen in the air and a thin layer of aluminium oxide forms on the surface of the aluminium. Water and air cannot pass through the layer, so the metal is protected from being corroded by them.

Any substance which can dissolve this oxide layer will cause the aluminium to corrode. Acid foods such as tomatoes react with the oxide layer, thus corroding the aluminium.

Aluminium is a very good conductor of heat and electricity. All the useful properties of aluminium mentioned, and its abundance on earth, make it a very important metal.

ACTIVITY 2.10 TESTING FOR ALUMINIUM

You are going to test for the presence of aluminium in a compound.

You will need 0.1M aluminium sulphate solution, dilute ammonia solution, dilute sodium hydroxide solution, two dropping pipettes, two test tubes and a test tube rack.

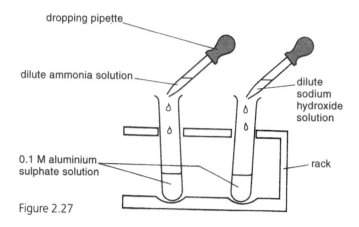

dropping pipette

dilute ammonia solution

dilute sodium hydroxide solution

0.1 M aluminium sulphate solution

rack

Figure 2.27

1 Place about 5 ml of the aluminium sulphate solution in each test tube.

2 To one test tube add six drops of ammonia solution, one drop at a time. Shake the test tube after adding each drop. Observe the solution carefully for any change after each drop is added.

3 Repeat step 2 with the second test tube, but this time use sodium hydroxide solution instead of ammonia solution.

4 Write down your observations in your note book. Write down how the aluminium compound behaved with the ammonia solution and with the sodium hydroxide solution.
Were there any similarities?
Were there any differences?

5 You are given a clear liquid compound called A, and want to find out if aluminium is one of the elements present in this compound.
Describe:

a what you would do

b the observations you expect if aluminium is present

c the observations you expect if aluminium is *not* present.

REACTIONS OF ALUMINIUM WITH ACIDS AND ALKALIS

(To be demonstrated by the teacher.)

You are going to be shown how aluminium reacts with hydrochloric acid and with sodium hydroxide solution.

You will need aluminium powder, two test tubes, a test tube rack, a wooden splint, two cork stoppers, dilute hydrochloric acid, concentrated hydrochloric acid, dilute sodium hydroxide solution and three dropping pipettes.

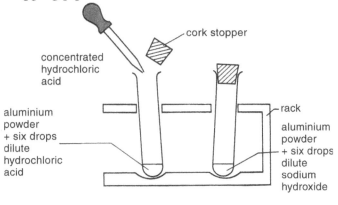

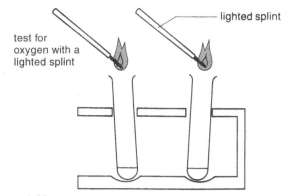

Figure 2.28

1. Put a spatula of aluminium powder in each test tube.

2. Add six drops of dilute hydrochloric acid to test tube A and put a cork stopper in the top.

3. Shake the tube gently. If there are no signs of reaction, carefully add six drops of concentrated hydrochloric acid to one tube. Replace the stopper, and feel the outside of the test tube.

4. To the other tube add six drops of dilute sodium hydroxide. Put a cork stopper in the top.

5. Test both tubes for gas using a lighted splint.

6. Record your observations in your notebook. What statement can you now make about the behaviour of aluminium with a) an acid and b) an alkali?

Aluminium and its alloys

Aluminium has many useful properties: it resists corrosion and it is a good conductor of heat and electricity. Aluminium is made stronger and easier to shape by mixing it with other metals to form alloys. One alloy of aluminium is Duralumin. It is a mixture of 1 per cent magnesium, 4 per cent copper and 95 per cent aluminium.

Aluminium and its alloys are used for making pots and pans, cameras, foil, window frames, vehicle bodies, electrical cable and aeroplanes.

Figure 2.29 *These things are made from aluminium*

Extracting aluminium from its ore

The bauxite ore is crushed and then purified to obtain aluminium oxide. The oxide is melted by mixing it with a salt called cryolite and heating the mixture to

850°C. An electric current is then passed through the melted oxide. The current causes the oxide to separate into its two elements: oxygen and aluminium. This process of separation is called **electrolysis**.

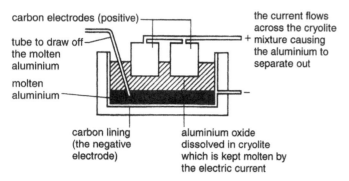

Figure 2.30 *Extracting aluminium by electrolysis*

Aluminium is extracted by electrolysis because it is a reactive metal, and it requires a large amount of energy to separate it from the other elements with which it combines.

When aluminium oxide is separated from bauxite, some of the impurities removed cannot be used. These waste products form red mud which is left in 'lakes' near bauxite mines. The red mud is strongly alkaline and harms plants and animals, but the problem of what to do with it has not yet been solved.

2.7 Compounds and mixtures

The importance of elements

Why do you think that elements are important? Well, as you have seen already, the many substances which make up living and non-living matter are formed from combinations of elements. Knowledge of elements helps us to understand the properties and possible uses of these other substances.

Two or more elements can combine chemically, or react together, to form a new substance that is completely different in appearance and properties from the individual elements and is called a **compound**.

To observe elements combining to form a compound, carry out the following activity.

ACTIVITY 2.12 MAKING A COMPOUND

You are going to investigate the chemical combination of two elements: sulphur and iron filings.

You will need 4 g of sulphur, 7 g of iron filings, a boiling tube, a Bunsen burner, a retort stand and clamp, and a mortar and pestle.

❶ Place the iron filings and sulphur in the mortar and grind until they are thoroughly mixed. Observe the mixture and record its colour and appearance. Put a magnet near the mixture and record your observations.

❷ Set up the apparatus as shown in the diagram and heat the mixture slowly until a red glow spreads throughout the mixture. Make sure the tube is slanted while its contents are being heated.

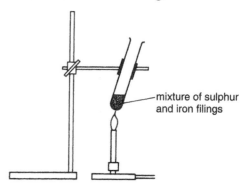

Figure 2.31

❸ Stop heating when the red glow spreads through the mixture.

❹ Allow the **residue** (what is left) in the tube to cool. When it is cool, examine it. Record what you observe. Put a magnet near the residue.

What happened when you put a magnet near the mixture of iron and sulphur before heating?

What happened when you put a magnet near the residue formed after heating?

❺ List the differences between the mixture before it is heated, and the residue formed after heating.

Is the residue the same colour as the mixture before heating? Is the residue powdery?

The residue is quite different from the mixed sulphur and iron filings. It is a new substance – a compound – called iron sulphide.

The red glow that spread through the mixture when it was being heated was caused by the two elements combining chemically.

Iron sulphide is made of two elements only, but many compounds are made of more than two elements. Here is a list of some well known compounds and the elements of which they are made. Find out some of the differences between the compounds and the elements which make them up.

Compound	Elements which make up the compound
water	hydrogen and oxygen
sodium chloride (common salt)	sodium and chlorine
cane sugar	carbon, hydrogen and oxygen
protein (e.g. meat)	carbon, hydrogen, oxygen, nitrogen and (usually) sulphur
carbon dioxide	carbon and oxygen
sulphuric acid	hydrogen, sulphur and oxygen
nitric acid	hydrogen, nitrogen and oxygen

Mixtures

A mixture is made up of two or more substances, but these substances are not chemically combined so each substance does not lose its individual properties.

A mixture can be made up of two or more elements. In Activity 2.12, when the sulphur and iron filings were ground together in the mortar, a mixture was formed.

Mixtures can also be made when two or more compounds are put together. For example, when sugar (a compound) dissolves in water (a compound) a sugar **solution** is formed. The sugar solution is a mixture.

A suspension of soil in water is also a mixture. A **suspension** is a mixture in which particles of one substance are found throughout a liquid or a gas; the particles do not sink to the bottom or rise to the top, but remain **suspended** in the liquid or gas.

Air is a mixture of gases; sea water is a mixture of water, sodium chloride and other substances. Soil and milk are also mixtures.

Alloys are mixtures of two or more metals. Bronze is an alloy; it is a mixture of copper and tin. Brass is another alloy; it is a mixture of copper and zinc.

Mixtures can be separated into their individual **components** (parts) by simple methods such as filtration, distillation and settling.

Comparing compounds and mixtures

Compound
Made when two or more elements combine chemically.
Not easy to separate into its individual elements.
Different in appearance and properties from its individual elements.

Mixture
Made when elements or compounds combine by physical means, for example, by mixing.
Can separate into individual elements or compounds by simple means.
Has the same properties as its individual components do when they are separate.

Figure 2.32 *A suspension of clay particles in water*

ACTIVITY 2.13 SEPARATING A MIXTURE

You are going to find out if the black ink of a felt tip marker is made of one colour of ink or a mixture of colours.

You will need a black felt tip marker, Petri dish, filter paper, a dropper and water.

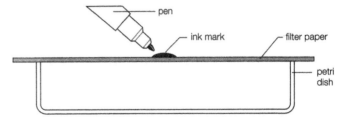

Figure 2.33

1 Make an ink mark in the middle of the filter paper with the felt tip pen.

2 Place the filter paper on the Petri dish.

Place two or three drops of water on the ink spot.

④ Write down your observations. Does the ink consist of more than one colour?
This method of separating mixtures of different colours is called **paper chromatography**.

2.8 Everyday acids

Acids are very important compounds. They are found in many things we use. For example, fruits which have a sharp, sour taste contain acids. Limes, oranges, grapefruit and tangerines contain **citric acid**.

Vinegar, which we use for pickling fruits and vegetables, is a dilute solution of **ethanoic (acetic) acid**.

Milk goes sour because of an acid, called **lactic acid**, which is formed in the milk.

Many fruits and vegetables contain **ascorbic acid**, which is another name for vitamin C. The little red West Indian cherry is very rich in ascorbic acid.

When tea has been allowed to stand for a long time, it tastes bitter. This is because of the **tannic acid** present in tea leaves.

Our stomachs produce **hydrochloric acid**, which, when very dilute, helps in the digestion of food.

Figure 2.34 *Cherries and vinegar are acidic*

Care with acids

Acids burn the tissues of plants and animals. Concentrated acids are very dangerous chemicals. They can seriously damage living tissues and burn through clothes and eat into metals. They must be handled with great care when they are used in the laboratory.

The stings of ants contain an acid called **methanoic (formic) acid**. The burning you feel when an ant stings you is caused by the formic acid it injects into your skin. The acid irritates the skin and flesh. However, the damage caused by this acid is much less than the damage that would be caused by a concentrated acid.

2.9 Physical properties of acids

You will see acids in your school laboratory. Some examples are: sulphuric acid, hydrochloric acid, nitric acid and ethanoic acid. A car battery contains sulphuric acid: you must be careful not to spill it.

ACTIVITY 2.14 DESCRIBING LABORATORY ACIDS

You are going to describe the physical appearance of some acids in the school laboratory.

You will need reagent bottles containing dilute nitric acid, dilute sulphuric acid, dilute hydrochloric acid and ethanoic acid.

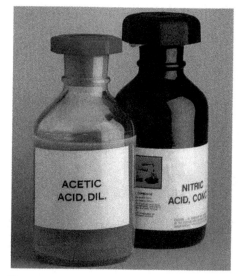

Figure 2.35

1. Look at each acid and describe its appearance:

Does it have a colour?
Does it have an odour?
What physical state is it in?

2. Use a chemistry textbook to find out what atoms are present in each of the acids.
Are acids elements or compounds? Give a reason for your answer.

The acids you have just observed are all liquids. But some acids exist as solids, for example ascorbic acid and citric acid. When you buy citric acid from the drug store, you receive a white crystalline solid. Vitamin C is usually sold in tablet form. The tablets behave like acids once they are dissolved in water.

Testing for acids

Very often you may have a liquid and not know whether it is an acid. In the activity that follows, you are going to learn a way of showing whether a liquid is an acid or not.

ACTIVITY 2.15 ACIDS AND INDICATORS

You are going to investigate the effect of acids on litmus, phenolphthalein and methyl orange (these are all called **indicators**).

You will need a set of clean test tubes in a rack, dropping pipettes, red and blue litmus paper, litmus solution, methyl orange, phenolphthalein, a watch glass, a lime, dilute sulphuric acid, dilute hydrochloric acid and dilute nitric acid, a knife.

1. Arrange four pieces of red litmus and four pieces of blue litmus on the watch glass as shown in Figure 2.36.

2. On one piece of red litmus paper place a drop of sulphuric acid, on the second place a drop of hydrochloric acid, on the third place a drop of nitric acid and on the fourth squeeze a drop of lime juice. Record your observations in your note book.

3. Repeat this procedure using the blue litmus paper instead of the red.

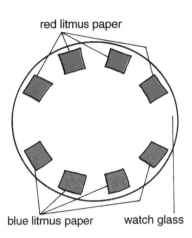

Figure 2.36

4. You are now going to carry out a similar procedure using the liquid indicators. First copy the table on page 36 into your note book. The table shows the effect of acids on indicators.

5. Note the colour of each of the indicators before you add them to the acids.

6. Follow the procedure shown in Figure 2.37, adding drops of each of the indicators to sulphuric acid in the test tubes. Note the colour of the resulting solution in your table. (Don't worry if you do not get a colour change with one of the indicators – you will see why we use this indicator later in the unit.)

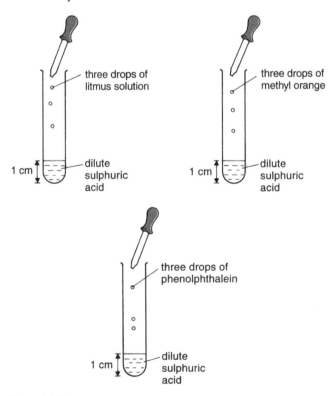

Figure 2.37

	Litmus solution	Methyl orange	Phenolphthalein
colour before acid added			
dilute sulphuric acid			
dilute hydrochloric acid			
dilute nitric acid			
lime juice			

7 Repeat these steps with each of the other acids. Record your results.

What inference can you make about the effect of acids on each of the indicators?

If you are given a clear colourless solution, how would you find out if it is an acid or not?

Explain what you would do and what you would expect to see if the solution is an acid.

Why do you think methyl orange, litmus and phenolphthalein are called indicators?

On your own

1 Use an indicator to test whether or not some local beverages are acid. You could test, for example, juices made from guava, plum, five finger or carambola, and passion fruit.

2 The dyes from some local flowers have been used as indicators. Using flowers available to you, try to determine if dyes extracted from them can be used as indicators for acids. Ask your teacher to help you in the extraction of dyes. Keep your dyes safely for use later on.

Litmus is a colouring or dye extracted from plants called lichens. A lichen consists of two types of plants – an alga and a fungus – living together. Lichens are found on rocks and trunks of trees. Litmus paper is absorbent paper soaked in the litmus dye. Litmus solution is the dye dissolved in a liquid solvent.

2.11 Chemical properties of acids

Substances in different groups have different properties. These properties include the ways in which they react with other substances. In this section, you are going to find out how acids react with two kinds of substances, namely metals and a group of compounds called carbonates.

ACTIVITY 2.18 ACIDS AND METALS

You are going to investigate the effect of acids on two metals: magnesium and zinc.

You will need a set of clean test tubes in a rack, a wooden splint, dilute hydrochloric and sulphuric acids, 1 cm of magnesium ribbon, some granulated zinc.

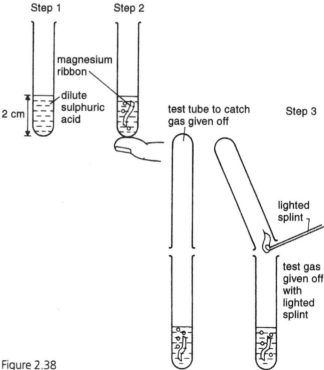

Figure 2.38

1 Pour dilute sulphuric acid into a test tube to a depth of about 2 cm. Drop a short piece of magnesium ribbon into the test tube.
What do you see?

2 You can see that the magnesium and acid are reacting because bubbles are being given off. With your other hand, touch the bottom of the test tube.
How does it feel?

3 Hold an empty test tube upside down over the test tube in which the reaction is taking place. After a little while, hold a lighted splint near the mouth of the upside down test tube.
What do you see?
What do you hear?

4 Repeat these steps using zinc instead of magnesium. Note which metal reacted more vigorously (was more **reactive**). Repeat the steps with both metals using hydrochloric acid. Answer the following questions:

a Which metal reacted more vigorously with the acids?

b The gas given off during the reaction is hydrogen. With which metal was hydrogen given off more rapidly?

c What test was used for hydrogen gas?

d Why was it necessary to use an upside down test tube to collect the hydrogen gas?

e From your observations in these activities, write down one main difference between magnesium and zinc.

f Copy these sentences into your book and fill in the blanks:
When metals react with acids, _____ gas is given off.
Some metals are more _____ than others.

ACTIVITY 2.17 ACIDS AND CARBONATES

You are going to investigate the effect of acids on carbonates and hydrogen carbonates.

You will need a set of clean test tubes in a rack, dilute sulphuric and hydrochloric acids, solid calcium carbonate, sodium carbonate, sodium hydrogen carbonate (baking soda), lime water in a test tube, bicarbonate indicator solution in a test tube, two spatulas and a one-holed stopper fitted with a delivery tube.

1 Carry out the steps shown in the diagram. Observe the reaction and record your observations.

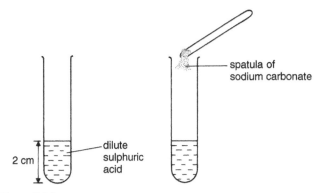

Figure 2.39

2 Repeat this procedure but, this time, quickly place the rubber stopper in the test tube and dip the end of the delivery tube in the lime water as shown below.
What happens to the lime water?

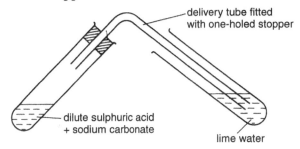

Figure 2.40

3 Repeat step 2 using:

a Dilute sulphuric acid and calcium carbonate; dilute sulphuric acid and sodium hydrogen carbonate

b Dilute hydrochloric acid and each carbonate in turn.

Record your observations in each case.

4 Repeat this activity, and use bicarbonate indicator solution instead of lime water.

The bubbles that are formed show that a gas is given off when an acid reacts with a carbonate. This gas is carbon dioxide.
What is the effect of carbon dioxide on lime water?

1 You are told that egg shells and baking powder contain carbonates. Carry out your own investigations to see if this statement is true.

2 Squeeze a lime and pour the juice into a cupful of water. Add a spoonful of baking powder. Can you explain your observations?

3 You can use what you have learnt to get a hard-boiled egg into a bottle whose neck is slightly smaller than the egg. Soak an egg in vinegar for several days. The shell becomes soft as the calcium carbonate is dissolved from the shell. Now you can push the egg into the bottle. To make the egg hard again, pour bicarbonate of soda into the bottle and leave it for a day. Show it to your friends and see if they can explain how the egg got in there.

2.12 Bases

Another important group of compounds is the bases. Many bases are insoluble in water, but some dissolve in water. The ones which are soluble in water are called **alkalis**.

Ammonia solution, milk of magnesia and caustic soda are examples of bases we use in our homes. Caustic soda, or sodium hydroxide, is used to make soap. Milk of magnesia contains the base magnesium hydroxide. It is used to soothe the stomach when you have indigestion because the contents of the stomach have become too acidic.

The lime water you used in Activity 2.17 to test for carbon dioxide is a solution of the base calcium hydroxide.

The word 'alkali' comes from the Arabic word for ashes, because alkalis were once prepared from the ashes of plants and wood.

Care with alkalis

Concentrated solutions of alkalis feel soapy and slippery. They also burn the skin, so you must handle them

Figure 2.41 *Alkalis are used in the manufacture of these products*

with care. If you ever spill a concentrated acid or alkali on your skin by mistake, quickly run water over the part that you have burned. This will dilute the acid or alkali and help to reduce the burning. If the burning is serious you should go to see a health worker.

2.13 Testing for alkalis

Can you remember how to test whether or not a liquid is an acid? You are now going to test whether or not a liquid is an alkali.

ACTIVITY 2.18 ALKALIS AND INDICATORS

You are going to investigate the effect of alkalis on the indicators you used in Activity 2.15.

You will need a set of clean test tubes in a rack, dropping pipettes, a watch glass, red and blue litmus paper, litmus solution, methyl orange, phenolphthalein, dilute sodium hydroxide solution and dilute potassium hydroxide solution.

1. Repeat all the steps you carried out in Activity 2.15, but this time using the sodium hydroxide and potassium hydroxide solutions instead of the acids.

2. Draw a table in your note book to record the results. What inference can you make about the effect of alkalis on each of the indicators used?

2.14 Strength of acids and alkalis

Not all acids and alkalis have the same strength. Hydrochloric acid is a stronger acid than citric acid;

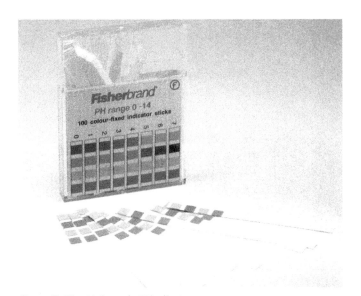

Figure 2.42   *Universal pH indicators*

sodium hydroxide is a stronger alkali than milk of magnesia. We can get a rough measure of how strong an acid or alkali is by using either pH paper (also called universal indicator paper) or universal indicator solution. You should be given a colour chart when you use these. This shows a range of colours with a number from 0 to 14 by each colour. The numbers are called the pH scale.

On your own

1. Copy this table into your note book and fill it in:

	Colour before acid or alkali added	Colour after acid added	Colour after alkali added
blue litmus paper red litmus paper litmus solution methyl orange phenolphthalein			

2. If you were given a clear, colourless solution, and you were asked to find out if it was an alkali, explain fully what you would do.

3. Use an indicator to test whether the following are alkaline: soda water, household bleach, liquid detergent, shampoo.

4. Find out whether the dyes you extracted from flowers on page 36 can be used as indicators for alkalis.

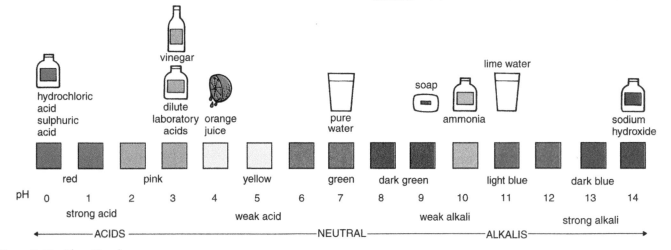

Figure 2.43   *The pH scale*

You are going to investigate how the colour of pH paper changes when you put it in various solutions.

You will need four clean test tubes in a rack, a watch glass, dropping pipettes, four pieces of pH paper, a pH colour chart, some soap solution, dilute sodium hydroxide solution and dilute hydrochloric acid.

1 Squeeze some lemon juice into one of the test tubes. Pour some soap solution into the second test tube, some dilute sodium hydroxide solution into the third and some dilute hydrochloric acid into the fourth.

2 Arrange the four pieces of pH paper on the watch glass.

3 With a dropping pipette, put one drop of the lemon juice onto one of the pieces of pH paper. Does it change colour?

4 Take the pH paper, and try to match the colour with the colours on the chart. Write down the number of the colour on the chart which is closest to the colour of the pH paper.

5 Repeat steps 3 and 4 with each of the other three solutions.

6 Arrange the four solutions in the order of the numbers you have written down. Which has the lowest number? Which has the highest number?

pH numbers of acids

The pH scale ranges from 0 to 14. Numbers 0 to 6 have colours ranging from red to orange. Acid solutions cause pH paper to show these colours. The lower the number, the stronger is the acid. You should have found that the lemon juice and the hydrochloric acid gave pH numbers of between 0 and 6.

A solution with pH 1 makes the pH paper become red. A solution with pH 5 makes the pH paper become orange. Both solutions are acids, but the solution of pH 1 is a stronger acid than the solution of pH 5. Which was the stronger acid: lemon juice or hydrochloric acid?

pH numbers of alkalis

Numbers on the pH scale from 8 to 14 have colours ranging from green to dark blue. Alkali solutions cause pH paper to show these colours. The higher the number, the stronger is the alkali. You should have found that the soap solution and the sodium hydroxide solution gave pH numbers of between 8 and 14.

A solution of pH 8 makes the pH paper become green. A solution with pH 14 makes the pH paper become dark blue. The solution whose pH is 14 is a stronger alkali than the solution whose pH is 8. Which was the stronger alkali: sodium hydroxide solution or soap solution?

Neutral solutions

Solutions that are neither acids nor alkalis are called **neutral**. A neutral solution has a pH of 7.

On your own

1 Find the pH of distilled water and of tap water. Are they neutral, acidic or alkaline?

2 Place the words 'greater than', 'less than', 'lower' or 'higher' in the blank spaces.
Acids have pH numbers _____ 7.
Alkalis have pH numbers _____ 7.
The _____ the pH, the stronger the alkali.
The _____ the pH, the stronger the acid.

3 Find the pH of several household substances.

2.15 Acid–alkali interaction

You are going to use an indicator to investigate what happens when an acid and an alkali interact.

You will need a 150 ml beaker, universal indicator solution and its colour chart, a 5 ml syringe, a 10 ml measuring cylinder, a glass rod, 20 ml of 0.1M sodium hydroxide solution and 20 ml of 0.1M hydrochloric

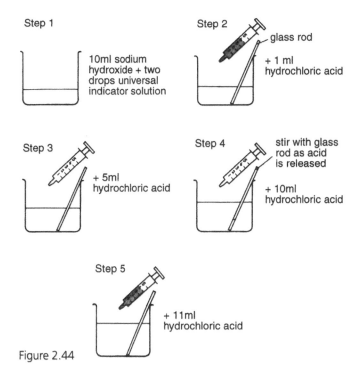

Step 1

10ml sodium hydroxide + two drops universal indicator solution

Step 2

glass rod

+ 1 ml hydrochloric acid

Step 3

+ 5ml hydrochloric acid

Step 4

stir with glass rod as acid is released

+ 10ml hydrochloric acid

Step 5

+ 11ml hydrochloric acid

Figure 2.44

acid. The abbreviation 0.1M shows how dilute the acid and alkali are. Your teacher will give you the acid and alkali you should use.

Use the diagrams in Figure 2.44 to help you carry out the activity.

1 Measure 10 ml of sodium hydroxide solution and pour it into the beaker. Add two drops of indicator solution. Find the pH of the solution and record it in your book.

2 Draw up into the syringe 5 ml of acid. Release 1 ml of the acid into the beaker with the alkali and mix it with the glass rod.
Is there any colour change? Record the pH in your book.
Is the solution acid or alkaline?

3 Add the rest of the acid in the syringe to the alkali. Again note the pH. Has it changed?

4 Draw up into the syringe another 5 ml of acid. Slowly release it into the beaker, stirring as you do so. Do you notice any change?
Record the pH.
Is the solution acid or alkaline?

5 Again, draw 5 ml of acid into the syringe. Release 1 ml of the acid into the beaker. Do you see any colour change?
Record the pH.
Is the solution acid or alkaline?

6 Release the remaining acid into the beaker. Does the colour change?
Is the solution acid or alkaline?

Figure 2.45a

Wasp stings are alkaline. If you put vinegar on a wasp sting, it neutralizes the alkali and helps to reduce the pain.

Figure 2.45b

Bee stings are acidic. Put baking powder, a weak alkali, on a bee sting to neutralize it.

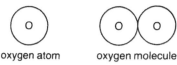

Acids and alkalis neutralize each other when they are mixed and new substances called salts are formed. Water is formed at the same time.

In the above activity, hydrochloric acid and sodium hydroxide reacted to form sodium chloride and water. Sodium chloride is common salt. If you had used a different acid or a different alkali, a different salt would have been formed.

2.16 Symbols and formulae

Formulae of elements

You have already learnt that each element has a symbol. For example, the symbol for oxygen is O. But oxygen exists as molecules made up of two atoms each. This information is written in this way: O_2.

O_2 is called the **formula** of the oxygen molecule. The plural of 'formula' is **formulae**.

oxygen atom   oxygen molecule

Figure 2.46

The table below shows the formulae of elements existing as molecules of two atoms.

Element	Symbol	Formula of molecule
hydrogen	H	H_2
nitrogen	N	N_2
fluorine	F	F_2
chlorine	Cl	Cl_2
bromine	Br	Br_2
iodine	I	I_2

For elements that do not exist as molecules their formulae are the same as their symbols. For example, Cu is the symbol for copper; it is also the formula for a molecule of copper. Some of these elements are shown below.

Element	Symbol	Formula of molecule
calcium	Ca	Ca
carbon	C	C
iron	Fe	Fe
neon	Ne	Ne
sulphur	S	S
zinc	Zn	Zn

Formulae of compounds

Compounds also have formulae. The formula for a molecule of water is H_2O. This tells us that each molecule of water is made of two hydrogen atoms combined with one oxygen atom.

Figure 2.47   water molecules

The table below shows some compounds and their formulae.

Compound	Formula of molecule
ammonia	NH_3
carbon dioxide	CO_2
methane	CH_4
glucose	$C_6H_{12}O_6$

For each of the compounds listed in the above table, write down how many of which types of atom are contained in one molecule of the compound.

2.17 Grouping elements into families

You have now looked in detail at three of the 106 known elements. Most of the 106 elements are solids, some are gases and two are liquids. These elements can also be grouped as metals and non-metals, or as elements that are coloured and elements that are colourless. How can these many elements be arranged so that it is easy to study them and the compounds which they form? Well, scientists solved this problem by grouping elements in a table called 'the periodic table of elements'.

Early attempts to classify elements

The search for ways to group elements according to some kind of pattern started in the early nineteenth century. One of the scientists who made a contribution to the grouping of elements was John Newlands, an Englishman.

In 1864, around 60 elements were known. Dalton had already shown that the atoms of these elements had different weights. Newlands arranged these elements in order of increasing atomic weight and drew up a table that arranged the known elements into seven 'octaves'. This table is shown below.

1	2	3	4	5	6	7
H 1	Li 2	Be 3	B 4	C 5	N 6	O 7
F 8	Na 9	Mg 10	Al 11	Si 12	P 13	S 14
Cl 15	K 16	Ca 17	Cr 18	Ti 19	Mn 20	Fe 21
Co, Ni 22	Cu 23	Zn 24	Y 25	In 26	As 27	Se 28
Br 29	Rb 30	Sr 31	Ce, La 32	Zr 33	Di, Mo 34	Ro, Ru 35
Pd 36	Ag 37	Cd 38	Sn 39	U 40	Sb 41	Te 42
I 43	Cs 44	Ba, V 45	Ta 46	W 47	Nb 48	Au 49
Pt, Ir 50	Os 51	Hg 52	Tl 53	Pb 54	Bi 55	Th 56

Newlands' table had seven vertical columns. From his table he noted two important patterns:

a   Elements in the same vertical column had similar

properties (for example lithium, sodium and potassium in column 2 behave in similar ways).

b The eighth element, starting from any given one, is a kind of repetition of the first. This is similar to the notes in music, which are arranged in **octaves**. Newlands therefore called this repeating of similar properties the 'Law of Octaves'.

But there were three main problems with Newlands' grouping of elements:

a There was no place in this arrangement for any new element likely to be discovered.

b Some elements seemed misplaced. For example, iron in position 21 was in the same vertical column as sulphur in position 14. Clearly, iron and sulphur do not resemble each other in properties.

c There were places where two elements occupied the same position, for example, cobalt and nickel are both in position 22.

On your own

1 Examine Newlands' table and find sulphur and iron. In which vertical column are they?

2 Which positions in Newlands' table have two elements?

Mendeleev's periodic table

About six years after Newlands' proposal, Dmitri Mendeleev, a Russian scientist, arranged elements in a table. The present-day periodic table is based on this grouping of the elements. Mendeleev's periodic table was published in 1871.

Mendeleev used the formulae of compounds to place the elements into groups. For example, those metals which he placed in Group 1 had the same general formulae in the chlorides and oxides which they formed.

He also arranged virtually all of the elements in order of increasing atomic weight.

On your own

Examine Mendeleev's table carefully. Spaces are left for elements with atomic weights 44, 68, 72, 100. Find these spaces. In which group is each space found?

Mendeleev called his grouping of elements the 'periodic table' because elements with similar chemical properties were repeated in the table 'periodically'.

Mendeleev's table consisted of eight elements running from left to right. They made up the **row** or **period**. The columns running from top to bottom were

Row	Group I	Group II	Group III	Group IV	Group V	Group VI	Group VII	Group VIII
1	H = 1							
2	Li = 7	Be = 9.4	B = 11	C = 12	N = 14	O = 16	F = 19	
3	Na = 23	Mg = 24	Al = 27.3	Si = 28	P = 31	S = 32	Cl = 35.5	
4	K = 39	Ca = 40	= 44	Ti = 48	V = 51	Cr = 52	Mn = 55	Fe = 56, Co = 59 Ni = 59, Cu = 63
5		Zn = 65	= 68	= 72	As = 75	Se = 78	Br = 80	
6	Rb = 85	Sr = 87	Yt = 88	Zr = 90	Nb = 94	Mo = 96	= 100	Ru = 104, Rh = 104 Pd = 106, Ag = 108
7		Cd = 112	In = 113	Sn = 118	Sb = 122	Te = 125	I = 127	
8	Cs = 133	Ba = 137	Di = 138	Ce = 140				
9								
10			Er = 178	La = 180	Ta = 182	W = 184		Os = 195, Ir = 197 Pt = 198, Au = 199
11		Hg = 200	Tl = 204	Pb = 207	Bi = 208			
12				Th = 231	U = 240			

called groups. All elements in the same group have similar chemical properties.

There were spaces in Mendeleev's table. These spaces, he said, were for elements which were not known at the time. Mendeleev was able to predict correctly the atomic weights and some chemical properties of these unknown elements.

The modern periodic table

The modern periodic table which we now use in our study of elements has the elements arranged in order of increasing atomic number instead of increasing atomic weight. The resulting ordering of the elements is very similar. You will learn about **atomic number** in the next section. This system of ordering elements has been accepted by scientists throughout the world.

The importance of the periodic table

The periodic table is a very important tool in chemistry. The position of an element in this table tells us several important things about the element. For example, it tells us whether it is a metal or non-metal, the other elements which are similar in chemical properties to it, and the kinds of compounds it will form.

On your own

1 Look at the chart of the modern periodic table in Figure 2.48.

   a How many elements are there in this table?

   b How many rows or periods does it have?

   c How many groups does it have?

2 Name the period and group to which these elements belong:

   a carbon   g oxygen
   b sodium   h iodine
   c aluminium   i sulphur
   d silicon   j helium
   e neon   k nitrogen
   f potassium

Pair off those elements in the list above which belong to the same group, and therefore have similar chemical properties.

3 Find hydrogen on the periodic table. Does hydrogen belong to any group?

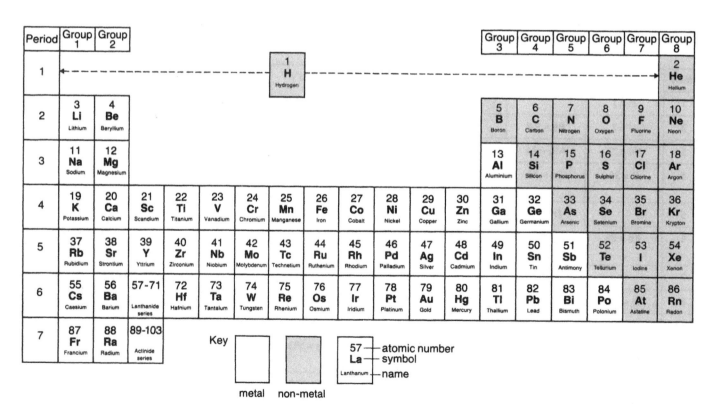

Figure 2.48 *The modern periodic table of elements*

Newlands' and Mendeleev's tables did not include the elements found in Group 8 of the modern periodic table. This group contains the inert gases, which are unreactive. They do not react with other elements so it was a long time before they were discovered. Newlands and Mendeleev did not know they existed.

Atomic structure

You have learnt, so far, that:

1 matter is made up of tiny particles called atoms; and

2 elements are substances each of which is made up of its own kind of atom.

What does an atom look like? Atoms are too small for scientists to see even with the most powerful microscope. So over the years, scientists have made mental pictures and physical **models** of atoms. Changes have been made to the models as the scientists gathered new information. Actually, scientists are still finding out more about the behaviour and structure of atoms. So it is possible that the present-day model of the atom may change.

A model of the atom

Many scientists have contributed to our present-day knowledge of the structure and behaviour of the atom.

Figure 2.51 on page 46 gives the names of some of these scientists, what they proposed, diagrams of their models of an atom, and the year/years in which their proposals were made.

Scientists' understanding of atoms is growing all the time. The model we use can be more or less complicated depending on what we want to use it for. (Generally, a model gives us an idea of what something is like.) The model of an atom described below is a useful one for understanding how elements behave and how they interact with each other.

An atom of any element consists of a very small **nucleus** which is surrounded mostly by empty space. **Electrons** move very rapidly in this empty space. The nucleus consists of **positively charged** particles, called **protons**, and particles with **no charge**, called **neutrons**.

The electrons are tiny **negatively charged** particles. They move rapidly around the nucleus in **shells** or **energy levels**. The electrons move so rapidly that they form an electron cloud.

A force of attraction exists between particles which have opposite charges. So there is a force of attraction between the nucleus and the electrons which move around it.

An atom has the same number of electrons as protons. A proton has one positive charge and an electron has one negative charge. In any atom the positive charges of the protons balance the negative charges of the electrons, so we say an atom is electrically **neutral**. That is, overall an atom has no charge.

An atom of the element oxygen. The oxygen atom has eight electrons (negative charges). The nucleus has eight protons (positive charges) and eight neutrons (no charge).

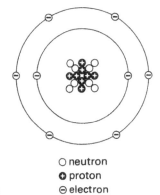

○ neutron
⊕ proton
⊖ electron

Figure 2.49

An atom of the element aluminium (Al). The aluminium atom has 13 electrons (negative charges). The nucleus has 13 protons (positive charges) and 14 neutrons (no charge).

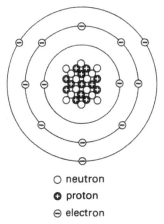

○ neutron
⊕ proton
⊖ electron

Figure 2.50

Scientists	What they proposed	Diagrams of models
John Dalton 1803 Manchester, England	All matter is made up of atoms. Each element is made up of its own kind of atom. Atoms are solid spheres.	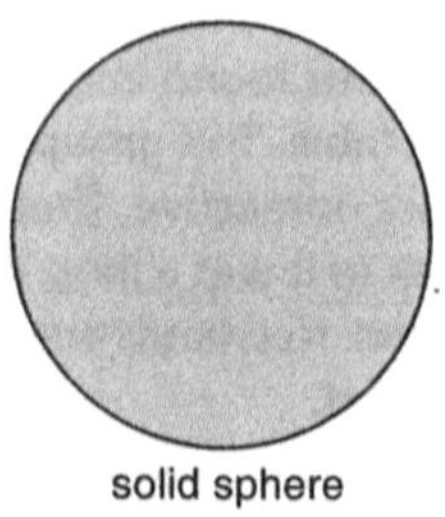 solid sphere
Joseph J Thomson 1897–1903 Cavendish Laboratory, Cambridge, England	Atoms have electrons. Electrons are negatively charged particles. Electrons of atoms are found in a ball of positively charged matter.	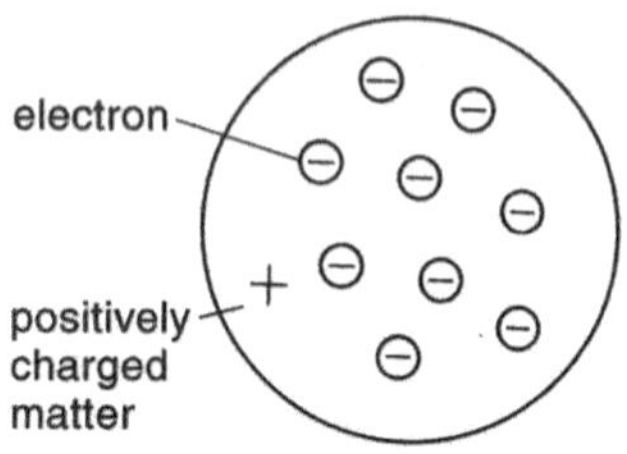 electron positively charged matter
Ernest Rutherford and Niels Bohr 1913 Cavendish Laboratory, Cambridge, England	An atom has a small, dense, positively charged nucleus. Electrons travel round the nucleus in orbits like planets orbiting the sun.	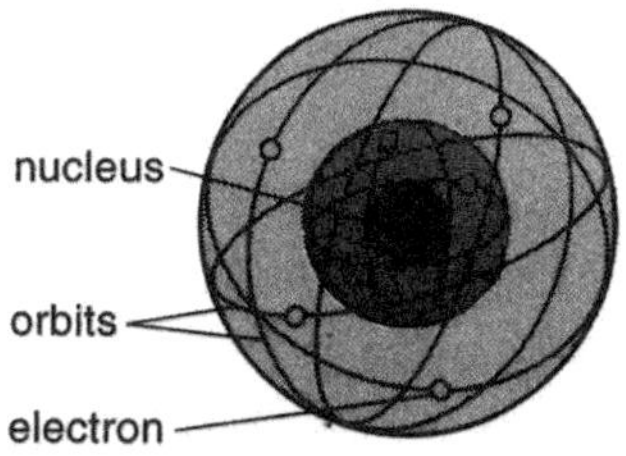 nucleus orbits electron
James Chadwick 1932	The nucleus of an atom contains positively charged particles (protons) and particles with no charge (neutrons).	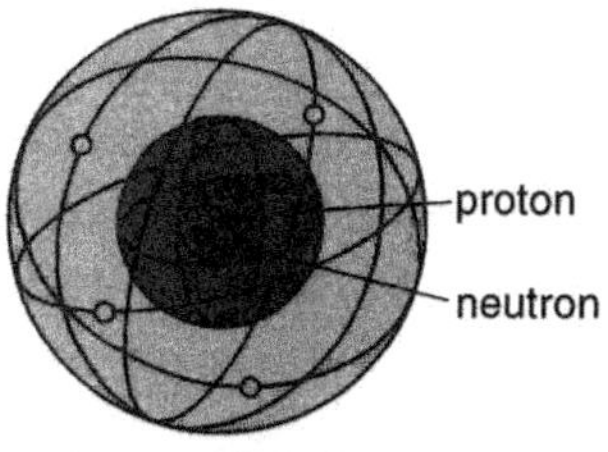 proton neutron
Many others up to the present day	The nucleus has many other particles besides neutrons and protons. Electrons move rapidly around the nucleus forming an 'electron cloud'.	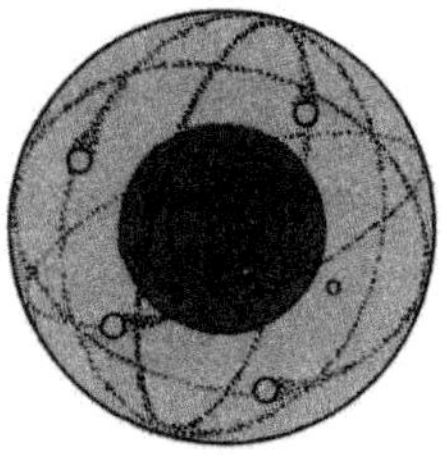

Figure 2.51   *Five different models of the atom*

Describing atomic structure

Three terms often used when describing atomic structure are mass number, atomic number and energy level.

The total number of protons and neutrons in the nucleus of an atom of any element gives the **mass number** of that element.

mass number = number of protons + number of neutrons

The mass of an atom is concentrated in the nucleus. The mass of the electrons is usually not included because it is very, very small and can be ignored.

The number of protons in the nucleus of an atom of an element gives the **atomic number** of the element.

atomic number = number of protons

Each element has the same number of protons in all of its atoms. No two elements have the same atomic number.

Let us look at oxygen. Its atom has 8 protons in its nucleus. Its atomic number is therefore 8. Only the element oxygen has atomic number 8.

As you have already seen, electrons are arranged in shells, or energy levels, around the nucleus of an atom. Each energy level can only hold a certain number of electrons. The first energy level holds a maximum of two electrons. The second energy level holds a maximum of 8 electrons. The third energy level holds a maximum of 18 electrons. A lower energy level must be filled up before electrons move on to the next level.

Let us look at the example of aluminium. An aluminium atom has 13 protons and 14 neutrons in its nucleus. Therefore we can say the following about aluminium:

mass number = number of protons + number of neutrons = 27

atomic number = number of protons = 13

number of electrons in one atom = number of protons = 13

number of electrons in 1st shell = 2

number of electrons in 2nd shell = 8

number of electrons in 3rd shell = 3

On your own

1 An atom of iron has 26 protons and 30 neutrons. How many electrons does it have?

2 An atom of chlorine has 17 electrons. There are 18 neutrons in its nucleus. How many protons are there in the nucleus?

2.19 Solutions

If you put a teaspoon of sugar into a cup of water, the particles of sugar will gradually disappear so that you cannot see them any more. The sugar is still in the water, but the molecules of sugar have become separate from each other and are spread out through the water. The mixture is now a **solution**. We say that the sugar has **dissolved**. In this solution the water is called the **solvent** and the sugar is called the **solute**.

Here are some examples of common solutions:

lemonade made with sugar, lemon juice and water; syrups made with sugar, water and extracts of fruit; coffee made of coffee powder and water; washing-up liquid made with liquid detergent and water.

On your own

Copy this table, which shows the first 12 elements, into your note book. Complete the table from carbon to magnesium.

Element	Symbol	Atomic number	Mass number	Number of particles in nucleus		Total number of electrons	Number of electrons in each shell		
				protons	neutrons		1st shell	2nd shell	3rd shell
hydrogen	H	1	1	1	0	1	1		
helium	He	2	4	2	2	2	2		
lithium	Li	3	7	3	4	3	2	1	
beryllium	Be	4	9	4	5	4	2	2	
boron	B	5	11	5	6	5	2	3	
carbon	C	6	12						
nitrogen	N	7	14						
oxygen	O	8	16						
fluorine	F	9	19						
neon	Ne	10	20						
sodium	Na	11	23						
magnesium	Mg	12	24						

Many liquid solutions have water as their solvent and a solid as a solute. We usually think of solutions in terms of a solid dissolved in water. There are some solids, however, that do not dissolve in water but in other solvents such as ethanol (alcohol), kerosene, turpentine and tetrachloromethane (carbon tetrachloride).

ACTIVITY 2.21 DISSOLVING IODINE CRYSTALS

You are going to investigate whether water or rubbing alcohol is the better solvent for iodine crystals.

You will need iodine crystals, two test tubes labelled A and B, a spatula, water, rubbing alcohol and two cork stoppers.

1. Place two crystals of iodine in each test tube. (Do not use your fingers; use the spatula.)

2. Half fill test tube A with water, and half fill test tube B with rubbing alcohol.

3. Close each tube with the cork stopper and shake gently for two minutes.

4. Allow the test tubes to stand for five minutes.

ACTIVITY 2.22 COMPARING NAIL POLISH REMOVERS

You are going to investigate which of two brands of nail polish remover is more effective.

You will need two different brands of nail polish remover, a brightly coloured nail polish, cotton wool and a dropping pipette.

1. Put two coats of nail polish on each thumb nail of one member of the class and allow to dry for two minutes. (Make sure the nails are the same size.)

2. Make two balls of cotton wool of the same size and put six drops of one brand of nail polish remover on one of them.

3. Use the cotton wool ball to remove the nail polish from one thumb nail, counting the number of times the nail is rubbed to remove the polish completely.

4. Put six drops of the other brand of nail polish

remover on the second cotton wool ball. Use it to repeat step 3 on the other thumb nail.

Which brand removed the nail polish completely with fewer rubs?

Which is the solvent and which is the solute in this activity?

Why did you use thumbs with nails of equal size, the same number of drops of nail polish remover for each thumb, and cotton wool balls of the same size?

How would you explain your observations?

Other types of solution

Solutions are not always made from solid solutes dissolved in liquid solvents. Examples of other kinds of solutions are given in the table below:

Kind of solution	Example of solution	Solute	Solvent
gases dissolved in liquids	aerated drink	carbon dioxide	water
gases dissolved in solids	marshmallow	air	sugar paste
solids dissolved in solids	alloys, e.g. brass	copper	zinc
liquids dissolved in liquids	alcoholic beverages, e.g. rum	alcohol	water

Figure 2.52 *These drinks are all solutions of carbon dioxide in water*

Properties of solutions

When a solute and solvent are completely mixed to form a solution, then all parts of that solution have the same chemical composition and chemical properties. We say the solution is **homogeneous**. 'Homogeneous' means the same all the way through. The molecules of the solute are spread evenly throughout the solvent.

The solute and solvent of a solution do not separate when the solution is allowed to stand for a long time.

Solutions and living things

Water is sometimes called a 'universal solvent' because a large number of substances dissolve in it. All living things are made up of protoplasm which is made up of about 80 per cent water. Many substances are dissolved in the water of protoplasm.

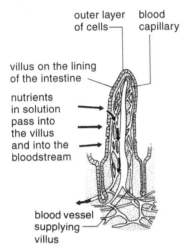

Figure 2.53   *Cross-section through a villus lining the intestine*

When animals digest food, the nutrients in the food are first dissolved. They can then be absorbed into the bloodstream from the digestive tract.

Plants take in nutrients from the soil in solution. When we add fertilizer to the soil, it dissolves in the water and the root hairs of plants take in the dissolved nutrients.

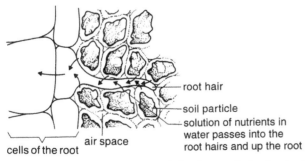

Figure 2.54   *Nutrients being taken in by the root hair of a plant*

Solutions and stains

Some substances make marks or stains on clothing. We can remove stains if we know which liquid solvents will dissolve them. The table below gives a list of substances that stain our clothes, and some possible solvents.

Substance	Solvent
grass stain	ethanol
chewing gum	tetrachloromethane
iodine	tetrachloromethane
nail polish	tetrachloromethane
oil paint	turpentine
tar	kerosene
grease	tetrachloromethane

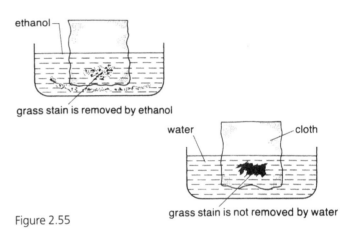

Figure 2.55

In the dry cleaning business tetrachloromethane is used to remove grease and fat stains from clothing. It is called 'dry cleaning' because the clothes are not cleaned with water.

On your own

Explain the following: to remove the green colouring (chlorophyll) from leaves, the leaves are crushed and ethanol added to them.

Concentrations of solutions

Are all solutions of one solute in one solvent the same? If you take sugar in your tea, how many teaspoons of sugar do you like? How can you tell the difference between a cup with one teaspoon of sugar and one with three?

ACTIVITY 2.23 DILUTE AND CONCENTRATED SOLUTIONS

You are going to make solutions of different concentrations.

You will need four beakers, water, some food colouring.

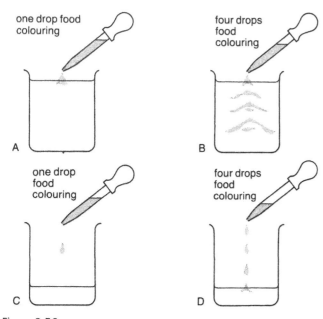

Figure 2.56

1 Pour water into the beakers and add drops of food colouring as shown in Figure 2.56.

2 Stir all of the solutions, and observe their colour.

Which solution has the deepest colour?

Which has the lightest colour?

How would you explain your observations?

The **concentration** of a solution depends on how much solute and how much solvent are used. If a solution has a small amount of solute and a large amount of solvent we say it is **dilute**.

When it has a large amount of solute and a small amount of solvent we say it is **concentrated**. We can change the concentration of solutions by changing either the amount of solute or the amount of solvent.

Saturated solutions

We can make a solution more and more concentrated by adding more solute. But with most solutions there comes a point when no more solute can be dissolved.

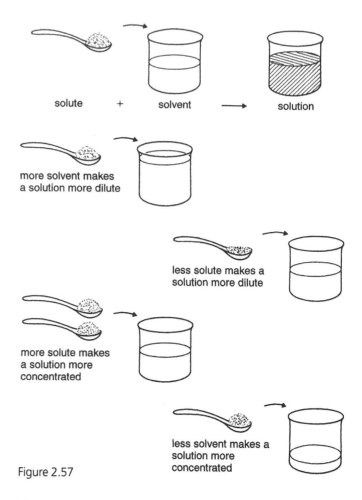

Figure 2.57

The excess solute settles out at the bottom of the solution. When this happens we say the solution is saturated.

Some solutions never become saturated. Ethanol can be dissolved in water up to any concentration and the ethanol will not settle out.

ACTIVITY 2.24 MAKING SATURATED SOLUTIONS OF SALT AND SUGAR

You are going to find out how much salt and how much sugar will dissolve in 100 cm³ of water at room temperature.

You will need two beakers labelled A and B, a measuring cylinder, water, sugar, salt, a beam balance, a spatula and a thermometer.

1. Measure 100 cm^3 of water into each of the beakers. Record the temperature.

2. Measure out 5 g of sugar, pour it into beaker A and stir.

3. Repeat step 2 until the solution is saturated. Record how much sugar you used.

4. Measure out 5 g of salt, pour it into beaker B and stir.

5. Repeat step 4 until the solution is saturated. Record how much salt you used.

Did more sugar or salt dissolve in 100 cm^3 of water at the given temperature?

In this activity, what variables were kept constant? What was the manipulated variable?

The amount of a solute which will dissolve in a solvent at a given temperature, before the solution becomes saturated, is called its **solubility**.

ACTIVITY 2.25 MAKING SATURATED SOLUTIONS AT DIFFERENT TEMPERATURES

You are going to compare the solubility of sugar in water at different temperatures.

You will need two beakers, sugar, water, a measuring spoon and a spatula.

1. Design an activity to measure how the solubility of sugar in water changes with temperature.

2. Draw a graph to show your results. You could put solubility on the vertical axis and temperature on the horizontal axis.

Does the solubility of sugar increase or decrease as the temperature increases?

How accurate do you think your measurements were?

How could you make them more accurate?

In this activity, what variables were kept constant? What was the manipulated variable?

The graph in Figure 2.58 shows the solubility of three substances in water at varying temperatures.

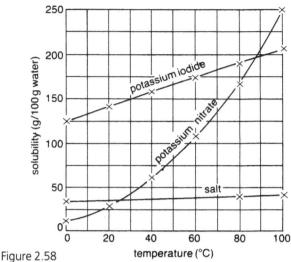

Figure 2.58

What can you say about the solubility of each substance as temperature increases?

Solutions as electrical conductors

Some solutions can conduct electricity. Electricity is passed through an aluminium oxide solution in cryolite during the extraction of aluminium from bauxite.

Many solutions of substances in water can also conduct electricity.

ACTIVITY 2.26 DO ALL WATER SOLUTIONS CONDUCT ELECTRICITY?

You are going to investigate whether solutions of substances in water can conduct electricity.

You will need four beakers, two electric cells, a bulb, two carbon electrodes, a switch, connecting wire and solutions of the following substances in water: sugar, cooking salt, copper sulphate and ethanol.

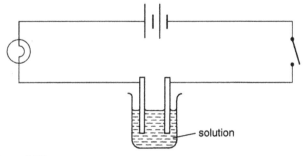

Figure 2.59

1. Set up the circuit as shown in the diagram above.

51

2 Put the electrodes into each of the four solutions in turn and close the switch.

Which solutions conducted electricity.

Which solutions did not conduct electricity?

This activity shows that water solutions can be grouped into those which conduct electricity, and those which do not.

A solution which conducts electricity is called an **electrolyte**. What is a non-electrolyte?

Three interesting chemical reactions

Are you aware that when a match bursts into flame, or a gas stove is lit, or iron rusts that chemical reactions are taking place? Let's take a close look at each of these three examples.

Matches

The coloured tip of a match is made up of three sets of substances. These substances are mixed with glue that holds them together. The substances are:

1 Compounds containing the elements phosphorus and sulphur. These compounds catch fire very easily.

2 A compound containing the element oxygen. This compound gives up its oxygen easily.

3 Powdered glass, which produces heat when rubbed.

The stick of the match is soaked in kerosene which helps the stick to burn easily.

The match box has a striking surface running along its two long sides. This surface is made of a mixture of powdered glass, a phosphorus compound and glue.

When the tip of the match is rubbed against the striking surface of the box, the heat produced by the rubbing surfaces causes a chain of chemical reactions to take place. First the oxygen is set free from the oxygen-containing compound. Then the freed oxygen reacts with the sulphur and phosphorus compounds. This reaction causes the tip of the match to burst into flame, which sets fire to the kerosene-soaked stick.

Safety matches will ignite only when rubbed on the striking surface of the correct type of box, or when placed near a flame or very hot surface.

Cooking gas

The gas we use for cooking in our homes can be either natural gas or liquefied petroleum gas, commonly called LPG.

In some countries, Barbados for example, natural gas is delivered to homes through pipes. It is obtained in the same way as petroleum: by drilling. It is often found where there is oil underground. Natural gas is mainly made up of a gas called methane.

Figure 2.61 *An offshore natural gas production platform, Trinidad*

Liquefied petroleum gas (LPG) is delivered to homes in gas cylinders. It is made up mainly of two gases – propane and butane – obtained as by-products in the refining of crude oil. These two gases are changed into liquid when placed under high pressure. It is in this liquid form that they are stored in cylinders.

Do you have a gas stove in your home? Is the gas delivered through pipes or in a cylinder? If your cooking gas comes in a pipe then you are using natural gas. If it comes in a gas cylinder you are using LPG.

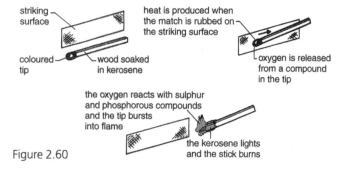

striking surface

coloured tip

wood soaked in kerosene

heat is produced when the match is rubbed on the striking surface

oxygen is released from a compound in the tip

the oxygen reacts with sulphur and phosphorous compounds and the tip bursts into flame

the kerosene lights and the stick burns

Figure 2.60

Burning of cooking gas

When cooking gas burns, a chemical reaction takes place. The compounds which make up the cooking gas contain hydrogen and carbon. Oxygen from the air reacts with the gas to produce heat; carbon dioxide and water are also formed. We can write the reaction in this way:

$$\text{gas} + \text{oxygen} \longrightarrow \text{carbon dioxide} + \text{water}$$

If the gas and air are in the right proportion, the gas burns with a blue flame. If there is too much gas for the amount of air, the flame is yellow, and soot (carbon) is formed. We can write the reaction in this way:

$$\text{gas} + \text{oxygen} \longrightarrow \text{carbon} + \text{water}$$

Advantages of using cooking gas

Using cooking gas has many advantages over wood or coal as a fuel. It is easy to light, it produces a clean flame and the amount to be used can be regulated on a stove. It is easily stored if it comes in containers.

One disadvantage is that it can escape and spread quickly. Since it can be lit easily, this can be quite dangerous when it escapes undetected.

If you can smell gas and think gas has escaped, NEVER light a match to see what has happened or turn on a light switch. Even a tiny spark in a light switch could make the gas explode. Instead, open all the windows to let the gas escape, and go and find help.

DANGER

Rusting

When metals are exposed to air, a chemical reaction takes place between the metal surface and substances in the air. Some of these substances are oxygen, water, carbon dioxide, sulphur dioxide and hydrogen sulphide.

When the chemical reaction takes place, a coating forms. This coating flakes off, exposing more metal surface to the air. This wearing away of metal surfaces caused by chemical action is called **corrosion**. Corrosion of iron or steel is called **rusting**. The coating formed is called **rust**. The chemical name for rust is iron oxide and its formula is Fe_2O_3.

Figure 2.62   *This old oil drum is coated in rust*

On your own

A student made the hypothesis that water alone causes iron to rust. To test this hypothesis he set up the investigation given below.
Investigation: Does water alone cause iron to rust?

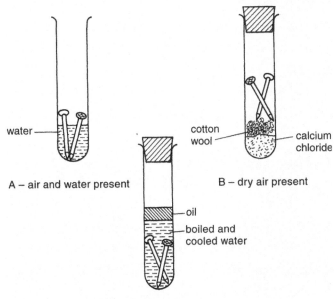

water

A – air and water present

cotton wool

calcium chloride

B – dry air present

oil

boiled and cooled water

Figure 2.63   C – only water present

Observations after a few days:
Tube A – the nails rusted. Some rust flaked off and was seen at the bottom of the tube.
Tube B – no rust.
Tube C – no rust.
Was his hypothesis correct?
What would you say causes iron to rust?
What was the calcium chloride in tube B used for?
Why was the water in tube C boiled?

Summary

These are some of the things you have learnt in this unit:

- All substances are made up of tiny particles called atoms.

- The particles in solids are closely packed and are arranged in an orderly pattern of rows and columns. The particles in a solid vibrate.

- The particles in a liquid can move over and past each other.

- Particles of gases are widely spaced and move quickly.

- Each element is made up of one kind of atom.

- Elements can be grouped into solids, liquids and gases, or they can be grouped into metals and non-metals.

- Many metals are malleable, ductile and have a shiny look, or lustre.

- Non-metals can be solids, liquids, or gases. They are brittle and poor conductors of heat and electricity.

- Mixtures can be made by physically combining elements and/or compounds.

- Compounds are formed when elements combine chemically.

- Mixtures can be separated by simple physical means. Compounds are not easily separated into individual elements.

- Acids and bases are two important groups of compounds. Bases which are soluble are called alkalis.

- Indicators are used to test a liquid to see if it is an acid or alkali.

- Acids have a pH less than 7 and alkalis have a pH greater than 7. A pH of 7 is neutral.

- An acid reacts with an alkali to form a salt and water. This type of reaction is called neutralization.

- Hydrogen is the lightest element. It is a non-metal. It is a constituent of water, all acids and many other compounds.

- The element carbon can exist in a number of forms and it is a constituent of fuels and all living things.

- Aluminium is the most abundant element in the Earth's crust and is extracted in five Caribbean countries.

- The formula of a compound shows how many atoms of which elements there are in each of its molecules.

- The periodic table arranges elements by increasing atomic number and helps us to make predictions about their properties.

- Atoms contain protons, neutrons and electrons.

- An atom is electrically neutral because it contains the same number of positively charged protons and negatively charged electrons.

- The mass number of an element is the number of protons and neutrons in the nucleus of one of its atoms.

- The atomic number of an element is the number of protons in the nucleus of one of its atoms.

- Electrons revolve around the nucleus in shells, or energy levels.

- In a solution a solute dissolves in a solvent.

- A solution has the same composition all the way through and all its parts have the same chemical properties.

- Water is very important to living things because of its ability to act as a solvent to many substances.

- The concentration of a solution is a measure of how much solute is dissolved in a given volume of solvent at a given temperature.

- A solution is saturated when no more solute will dissolve in it.

- Some solutions conduct electricity. They are called electrolytes.

- Matches contain substances which burn easily and compounds which give up oxygen.

- When gas burns it reacts with oxygen in the air and gives out heat energy.

- Corrosion occurs when metals react with substances in the air.

Q U E S T I O N S

A   Chromium   B   Carbon
C   Helium   D   Carbon dioxide

Use the list above to answer questions 1 and 2.

1 Which of these substances is *not* an element?

2 Which of these substances is an element and also a gas?

3 Which of the following statements is the best definition of an element? An element
   A   can be a solid, liquid or gas
   B   is a pure substance
   C   consists of one type of atom
   D   makes up natural substances

4 Which of these processes would you use to separate the different colours which make up chlorophyll extracted from a green leaf?
   A   Chromatography   B   Distillation
   C   Evaporation   D   Filtration

5 Give the symbol for each of the following elements:
   (i)   zinc
   (ii)   oxygen
   (iii) bromine
   (iv)   sodium
   (v)   potassium
   (vi)   silver
   (vii) tin
   (viii) nitrogen
   (ix)   helium

6 Give the names for the following elements:
   (i)   Cr
   (ii)   Co
   (iii) Cu
   (iv)   C
   (v)   Cl
   (vi)   Ca

7 (i) Imagine that you could cut up a lump of iron sulphide into the smallest piece that shows the properties of iron sulphide. What atom or atoms would that piece of iron sulphide contain? Is iron sulphide an element or a compound?

   (ii) Imagine you could cut up a lump of copper into the smallest piece that shows the properties of copper. What atom or atoms would that piece of copper contain? Is copper an element or a compound?

8 Draw two columns in your note book, name one column 'compound'; name the other 'mixture'. Place each of these substances in the correct column: air, water, brass, salt water, sodium chloride (cooking salt), sugar, sea water.

9 Copy this table in your note book and fill in the blank spaces.

Indicator	In acid	In alkali
phenolphthalein	colourless	——
methyl orange	——	yellow
litmus	——	——
universal indicator	orange to red	green to dark blue

10 a Copy the crossword puzzle below into your note book. Use the things you have learnt in this unit to solve the clues and complete the puzzle.

Figure 2.64

ACROSS
5 If a liquid turns litmus red it is _____ .
6 & 3 down An acid and an alkali react to give a _____ and _____ .
7 Given off when an acid reacts with a metal.
9 Hydrochloric acid is a _____ acid than lemon juice.
11 A more reactive metal than zinc.
12 pH paper turns this colour with a weak alkali.
13 _____ water turns cloudy with carbon dioxide.

DOWN
1 Some of these substances dissolve to form alkalis.
2 Turns yellow in alkali.
3 (See 6 across.)
4 An acid used in the home.
8 Another indicator.
9 Caustic soda is used in making this.
10 A solution with this pH number is a fairly strong alkali.

(The solution is given on page 179.)

10 b Copy this grid into your book. Circle the words, listed as (i) – (xv) below, in the grid.

```
M Y X V D A B B C J
O M U I C L A C O N
E M U E Q M O T P O
N E T I N U R G P B
I R L T S I L V E R
R C O X Q S M F R A
O U G X U E A O W C
L R S W Y N Z T R N
H Y D R O G E N O B
C N O G R A E F E P
S O D I U M A N S Y
```

   (i) argon
  (ii) bromine
 (iii) calcium
 (iv) carbon
  (v) chlorine
 (vi) copper
(vii) gold
(viii) hydrogen
 (ix) magnesium
  (x) mercury
 (xi) oxygen
(xii) potassium
(viii) silver
(xiv) sodium
 (xv) tin

(The solution is given on page 179.)

11

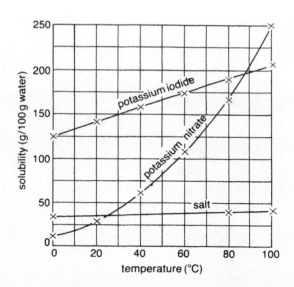

Use the graph above to answer these questions:

  (i) Name the independent variable and the dependent variable.
 (ii) Which solid has the highest solubility at room temperature?
(iii) Which solid has the lowest solubility at room temperature?
 (iv) Which solid dissolves most as temperature increases?
  (v) Which solid dissolves least as temperature increases?

12 Does baking powder contain a carbonate?
Design and carry out an experiment to find the answer to this question. Your design must include the following:

   (i) The aim of the experiment
  (ii) Hypothesis
 (iii) Materials
 (iv) Procedure
  (v) Your observations and result
 (vi) Explanation of result
(vii) Conclusion

Investigating forces

OBJECTIVES

- Understand what a force is and what it does, and be able to distinguish between different types of forces
- Perform simple experiments to investigate the relationship between force, mass and acceleration for a moving body
- Define and appreciate opposing forces as action and reaction
- Understand the definition of work and know the units used in its measurement
- Define and understand pressure and density

In this unit, you will investigate different forces, and show how some of these can be measured. You will find out how important some of these forces are in everyday life. You will also look at different types of machines and see how these can be useful.

Why is the ball moving in the photograph below? You could push or pull the ball to make it move. This is called a force. We cannot see a force, but we see its effects. A force can be exerted when two objects come in contact with each other as shown in the picture; or between two objects not in contact.

Figure 3.1 *Many forces operate during a football game*

3.1 Making objects move

Look at objects around you: are they moving? Things remain **stationary** (that is, not moving) unless something is done to them. Have you ever thought about what makes things move?

Balanced and unbalanced forces

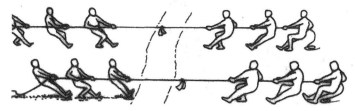

Figure 3.3

In a tug of war both teams are pulling, but in opposite directions. If they are pulling with the same force, neither team wins. What must a team do to pull the other team over the line?

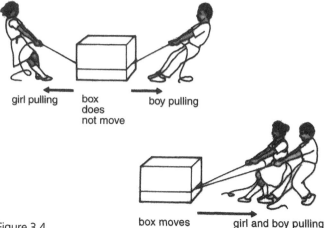

Figure 3.4

girl pulling   box does not move   boy pulling

box moves   girl and boy pulling

Look at the left-hand picture above. Explain why the box remains where it is. Why does it move when the boy and girl pull on the same side?

If two forces acting in opposite directions are **balanced**, they do not make a stationary object move. To make a stationary object move, one force must be greater than the other, that is they must be **unbalanced**.

Measuring forces

The strength of a force is measured in units called newtons (N) after the famous scientist Isaac Newton. We can measure the size of a force using a spring balance.

pulling force

spring balance

Figure 3.5

❷ Look at a spring balance. What is inside it? Why can a spring balance be used to measure the strength of forces?

❸ Get a spring balance with a scale that goes up to at least 100 N. Ask a friend to hold onto one end of the spring balance. With your little finger, pull on the other end. How big a force can you pull with? Can you pull with a bigger force using your index finger?

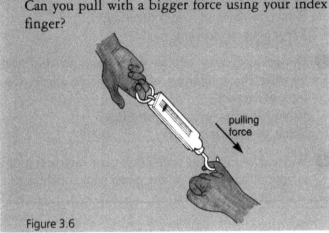

pulling force

Figure 3.6

3.2 Two important forces

The force of gravity

Why do you think that when you hold an object in the air and let go of it, it falls towards the ground? That is, it moves. There is a downward force on the object which makes it move. This force is due to the gravitational pull of the Earth on the object. We call this force **gravity**. All objects exert gravitational pull on each other, including us. However, we do not feel or see the effect of our gravitational pull and of other objects because we are all very small compared to the size of the Earth.

It wasn't my fault, Mum. It was gravity.

Figure 3.7

The downward force on an object due to Earth's gravitational pull is called its **weight**. We can measure the weight of an object using a spring balance.

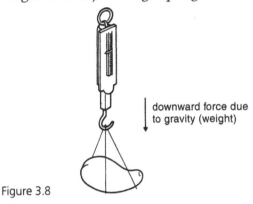

downward force due to gravity (weight)

Figure 3.8

On your own

Select a few objects in your classroom. Find the weight of each.

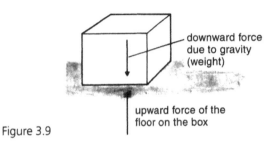

downward force due to gravity (weight)

upward force of the floor on the box

Figure 3.9

The box above remains at rest on the floor. Gravity is pulling the box downwards, so why is the box not moving?

When an object is at rest the forces on it are balanced. Let us see what forces act on this box. There is a downward force due to gravity. This force is the weight of the box. There has to be an equal upward force from the floor on the box. What would happen if the downward force was greater than the upward force?

Explain what forces are acting on a book which you are holding stationary in your hand.

Friction

You may have been worrying about a common observation: when we pull or push a stationary object, it doesn't move immediately. Have you ever tried to push a heavy box along the ground? Is it easy to make it move? Can you explain this?

You are going to measure the force needed to move an object.

You will need a spring balance, some string, a box and some sand.

1 Pour sand into the box so that it is quite heavy.

2 Tie string round the box as shown in the diagram.

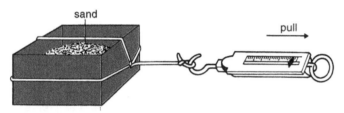

Figure 3.10

3 Using the spring balance, pull on the box until it just moves. Record the force required to make the box move.

The box does not move immediately. Before it moves, you are applying a force on the box in one direction, so why does it not move? There must be a force balancing it in the other direction. This force is called friction. Friction is a force which opposes motion.

Why does an object eventually move if you pull or push it hard enough?

Figure 3.11

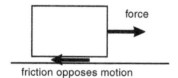 **ACTIVITY** **3.2** **INVESTIGATING FRICTION**

You are going to investigate the conditions that affect friction.

You will need a spring balance, boxes of different sizes, sand, string, surfaces of different roughness.

1 Design and carry out an experiment to find out whether the weight of an object affects the force of friction.
What variables are you controlling?
What are you changing?

2 Design and carry out an experiment to find out whether the area of an object touching a surface affects the force of friction.
What variables are you controlling?
What are you changing?

3 Design and carry out an experiment to find out whether the roughness of the surfaces affects the force of friction.
What variables are you controlling?
What are you changing?

4 Write down in your note book your interpretation of your observations. Write down your conclusions.

Friction can be a useful force. There is friction between the tyres of a bicycle and the ground. If there was no friction, the bicycle wheels would skid round and round without the bicycle moving forwards.

 # Objects in motion

Speed and velocity

To describe the motion of an object such as a car, we talk about its **speed**. The speed tells us how fast an object is going. That is, how much distance is covered in a certain time:

speed = distance/time

Speed is measured in metres per second – symbol m/s (or $m\,s^{-1}$).

For example, if a car travels 1 km in 100 seconds, then:

speed = 1000/100 = 10 m/s

It is sometimes more convenient to refer to speed using other units, such as kilometres per hour. But these can be converted into metres per second if necessary.

When we give the direction as well as the speed of a moving object, we are describing the **velocity** of the object.

Figure 3.12 *Runners running around a track*

Velocity gives **speed and direction**. For example, we can say that the velocity of a car is 80 kilometres per hour in a northerly direction.

Look at the picture above.

How does the athletes' speed change during the race? How does their velocity change?

Uniform and changing motion

The man shown in Figure 3.13 is travelling with uniform motion. What can you say about his speed? What about his direction? When an object travels in a straight line at constant speed the object is said to have **uniform motion**. We can also say that the object has **constant velocity**.

The scientist, Isaac Newton, observed objects in motion. He noticed that a change in motion takes place in certain situations. A ball changes its direction of motion when it bounces as shown in Figure 3.14.

When a car slows down, its motion is changing – becoming slower until it stops. What causes a change in motion?

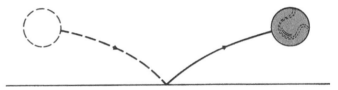

Figure 3.14

You have already seen that an unbalanced force can make a stationary object move. How else can a force affect the motion of an object? It can also change the direction of a moving object, stop a moving object, or change the speed of a moving object. That is, an unbalanced force can change the velocity, or motion, of an object. The illustrations in Figures 3.15a, b and c on page 62 show examples of unbalanced forces.

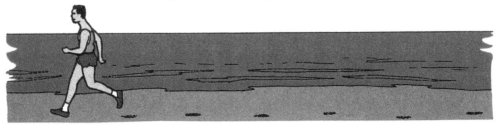

Figure 3.13

Figure 3.15a   *Why has this ball changed direction?*

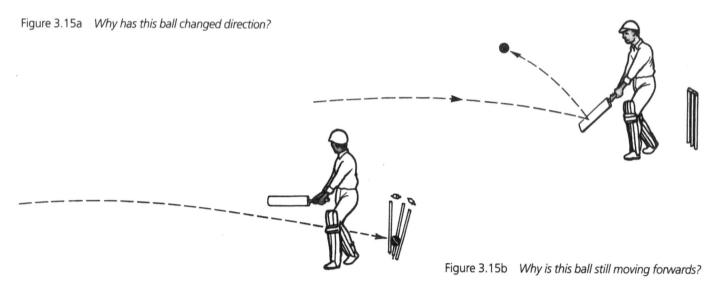

Figure 3.15b   *Why is this ball still moving forwards?*

Figure 3.16a   *Why has the car stopped?*

Figure 3.16b   *What is about to make the surfer move faster?*

Newton observed this and stated it as a law: **an object remains at rest or in uniform motion unless an unbalanced force acts on it.**

Describing changing motion

The velocity of an object changes if either its speed or its direction changes. There are ways of describing changes of velocity which take into account both speed and direction. Here we will talk only about changes in speed.

Imagine a car is being driven on a journey. It is stationary at the beginning of the journey. It speeds up, then the speed may become steady. The car slows down as it approaches crossroads and traffic lights, and speeds up again as it moves away from them. At the end of the journey, the car slows down until it is stationary.

When the car's speed is increasing, we say it is **accelerating.** The acceleration tells us the rate at which the speed is changing. For example, we can say that an object is accelerating at ten metres per second per second. This means that the speed of the object is increasing by ten metres per second in every second.

When the car is slowing down, we say it is **decelerating.** We can also say that it is showing **negative acceleration.**

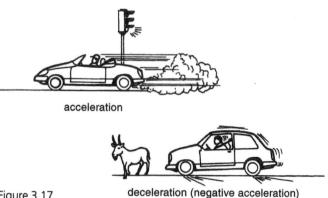

acceleration

deceleration (negative acceleration)

Figure 3.17

In each of the following situations, a change in motion takes place. Describe what the change in motion is, and what force causes the change.

1 The ball in a football match hits a goal post.

2 A seabird dives straight down into the water to catch a fish.

3 There is a sudden gust of wind behind a sailing boat.

4 You stop pedalling as you are cycling along a level road.

5 You have been free-wheeling on a bicycle along a level road, and you start going down a steep hill.

Relating force, mass and acceleration

If the unbalanced force acting on an object is big, how will this affect the acceleration of the object? For a given force, how does the mass of an object affect its acceleration?

You can investigate how speed changes when force and mass are changed.

ACTIVITY 3.3 FORCE, MASS AND ACCELERATION

You are going to investigate how force and mass affect changes in speed.

You will need a board or flat surface, similar type rubber bands, toy cars or toys on wheels. Use anything else you need.

1 Design experiments to find out (a) how speed changes when force is changed, and (b) how speed changes when mass is changed.
The design of your experiment is an important stage. For this investigation, you have to think of questions like these:
How will I check the speed?
What can I do to give toy cars a force?
When I change force, what must I keep the same?
What will I have to observe?

When I change mass, what must I keep the same?
What will I have to observe?

2 After thinking and designing, try out your design. You may want to improve it at this stage.
How will you change force?
How will you change mass?
How will you measure changes in speed?

3 As you do your experiments, record what you observe. You could do this in several ways. A table is one example. Often you make several observations.
Do your observations follow any pattern?
How does mass affect changes in speed?
How does force affect changes in speed?

Newton observed how force and mass affected acceleration (i.e. changes in speed). He found that:

1 when the mass is the same, the greater the force the greater is the acceleration produced; and

2 when the force is the same, the greater the mass the smaller is the acceleration produced.

3.4 Equal and opposite forces

Look at the pictures below.

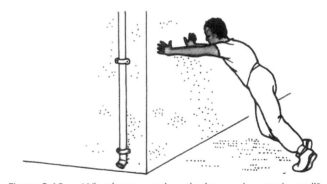

Figure 3.18a *What happens when the boy pushes on the wall?*

Figure 3.18b *This girl wants to move the boat forwards. How will she have to move her oar in the water?*

63

In these two cases, equal and opposite forces are taking place. The boy pushes with a force on the wall and there is an equal force from the wall pushing him backwards. The girl pushes the oar in one direction and the boat moves in the opposite direction.

For every force, there is an equal force in the opposite direction. We use this fact when we move. When you walk, your foot pushes backwards against the ground. There is an equal and opposite force on your foot which pushes you forwards.

Look at these pictures:

Figure 3.19

Sketch them in your note book, and show the equal and opposite forces. The equal and opposite forces are also called **action** and **reaction**.

Did you know that a rocket engine causes a rocket to move upwards by pushing gas downwards? The gases pushing downwards cause a reaction that moves the rocket upwards. Engines that work in this way are called reaction engines.

Energy and work

Look at the pictures below.

Figure 3.20

In each case a force is required – to lift the box, to push the car, to pull the cart. Also, in each case something moves. The box is lifted, the car moves and the cart moves.

All the people in the pictures are doing **work**. When we do work we use a force to cause movement. This is a special way of explaining work in science. If there is no movement then no work is done.

Figure 3.21 *No work is done!*

Work is the application of a force to move something over a distance.

work = force × distance

Force is measured in newtons and distance is measured in metres. Work is measured in newton metres (N m).

Work done in lifting objects

If you lift an object, you have to overcome the force of gravity. That is, you have to apply an upwards force which is at least equal to the weight of the object.

If the weight of the box is 5 N and you lift it to a height of 2 m the work done will be:

work = force × distance
     = 5 × 2 N m
     = 10 N m

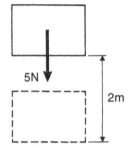

Figure 3.22

Work done against friction

Look at Figure 3.23. The people are pushing the car. What is the weight of the car? What is the force being used to push the car? The people are not lifting the car, so they do not need to overcome the force of gravity. However, they do need to overcome the force of friction. This force is acting between the wheels and the road.

If the car is pushed 100 metres, how much work is done?

work = force × distance
     = 600 × 100 N m
     = 60 000 N m

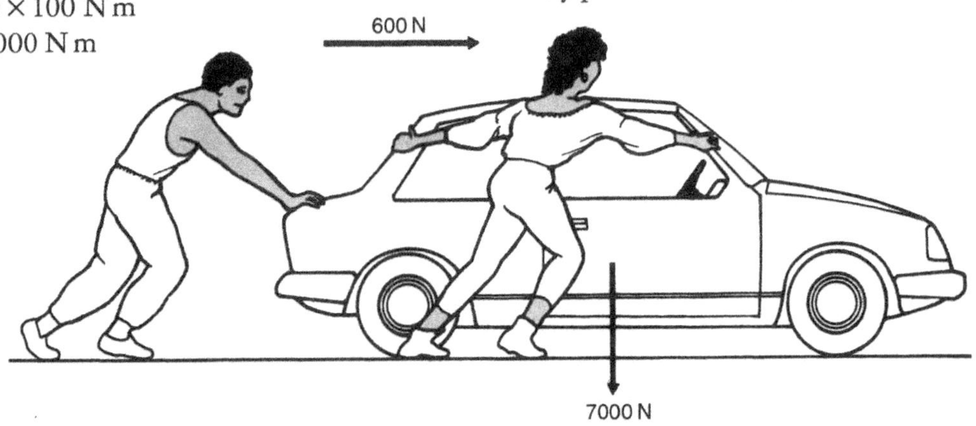

Figure 3.23

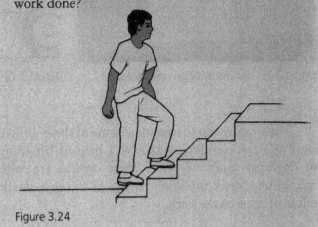

On your own

1 When this boy walks up the steps he is doing work. Explain why this is so. How can you calculate the work done?

Figure 3.24

2 A woman lifts a suitcase weighing 120 N up to a height of 50 cm. How much work was done? (Remember the unit for work is the newton metre.)

3.6 Machines

We need energy in order to do work. We do work when we apply a force and move something over a distance. We can also explain this by saying we apply a force to overcome a **resistance**.

The force we apply is called the **effort**. It is easy for us to lift a book, but we cannot in the same way lift a car. This is because we need to use a larger effort to lift a car than a book. Also, we should be careful when lifting heavy objects, since the size of the effort we need to use may put too much strain on our bodies.

Figure 3.25   *An ancient Egyptian pyramid*

Figure 3.26

What can we do to overcome some of these difficulties? We can use devices that will help us lift things or do other kinds of work. These devices are called **machines**. Many machines enable us to use a smaller effort in order to do work.

The inclined plane

A long time ago the Egyptians constructed some giant structures called the Pyramids (see Figure 3.25). Even today, people are amazed at how the Egyptians managed to build them. They did the work with the help of ramps. In the diagram above the men are able to lift the oil drums by rolling them up the ramp. (The ramp is a machine because it helps in doing the work. This type of machine is called an **inclined plane**.)

ACTIVITY **3.4** **USING AN INCLINED PLANE**

You are going to find how much force is needed to pull an object up an inclined plane.

You will need a spring balance, a stack of books or some blocks of wood, a board about 1 metre long and a toy car or any object that will roll on wheels.

1 Find the weight of the toy car.

2 Set up the inclined plane using the board and the books.

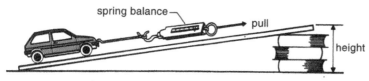

Figure 3.27

3 Measure the height of the inclined plane.

4 Pull the car along the length of the plane. What is the force required to pull the car?
Do this activity three times. Copy the table below in your note book and record your observations.

Trials	Length of plane (cm)	Height of plane (cm)	Weight of car (N)	Force needed to pull car (N)
1 2 3				

If you had to lift the car vertically upwards, what force would be required?
Compare this force with the force used when pulling the car up the plane.
What effect has the inclined plan had on the effort needed to raise the car to the same height?

Very often when you do investigations you have to take a number of 'readings'; that is, make your observations a few times. Why do you do this?

On your own

1 Design and carry out an experiment to investigate what force is required if you change the height of the plane.

2 Find what force is required if you increase the weight of the car.

In your activities you will have found that the force required to pull the car up the inclined plane is less than the force required to lift it vertically. But you don't get something for nothing. How far did you have to move the car when you lifted it vertically to the top of the board? How far did you move it when you pulled it along the board to the top? What can you say about the distance through which the object is moved using an inclined plane?

Work input and work output

If an object weighs 6 N, how much work will you need to do to lift it to a height of 0.5 m (see Figure 3.28)?

The force needed will be 6 N, so:

$$\text{work} = \text{force} \times \text{distance}$$
$$= 6 \times 0.5 \text{ N m}$$
$$= 3 \text{ N m}$$

By experiment it is found that to pull this object up the inclined plain requires a force of 4 N. The plane is 1 m long, and its top is 0.5 m high. Therefore:

$$\text{work done} = \text{force} \times \text{distance moved}$$
$$= 4 \times 1 \text{ N m}$$
$$= 4 \text{ N m}$$

The **work output** is the result of what happened. In this case, an object of 6 N was raised to a height of 0.5 m. The work output is 3 N m.

The **work input** is the work done in order to get the object to that height. When the object is pulled up the inclined plane, the work input is 4 N m.

The work input is usually greater than the work output because effort or force is necessary to overcome forces due to friction. Remember the force applied to do work is called the effort and the force overcome is called the resistance.

Another way of looking at this is in terms of energy. We use a certain amount of energy in raising the object to a higher level. This energy becomes potential energy in the object. We also use a certain amount of energy in overcoming friction. This energy is converted into heat energy which is created by two surfaces moving over each other.

If the work input is equal to the work output then the machine is said to be 100 per cent efficient. In the inclined plane and in other machines, if there was no friction then work input would equal work output.

The wedge

Figure 3.29

The **wedge** is a machine. It is a type of inclined plane. In your activities with an inclined plane, the plane remained in one place (it was stationary) and the object moved along the plane. What is moving in Figure 3.29? The wedge works like a double inclined plane. Where are the inclined planes in Figure 3.29?

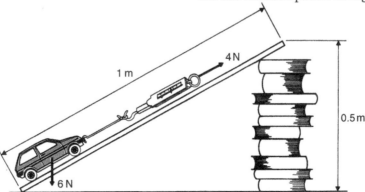

1 m

4 N

0.5 m

6 N

Figure 3.28

The lever

The hammer and the wheelbarrow shown in Figure 3.30 are being used as examples of machines called **levers**.

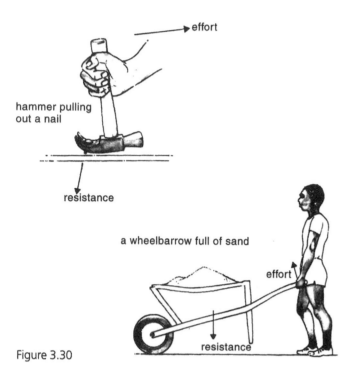

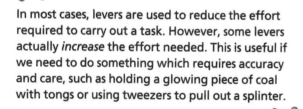

Figure 3.30

In both these diagrams one part of the lever is fixed. This part is called the **fulcrum**. Where is the fulcrum in each lever?

Levers are used to make work easier. If we tried to pull a nail out of a block of wood using our hands, we would have to use a very large effort. When a hammer is used to pull out a nail, a smaller effort can be used. Again, we don't get something for nothing. The nail we pull out moves through a very short distance. But our hand, which applies the effort, has to move through a greater distance.

In most cases, levers are used to reduce the effort required to carry out a task. However, some levers actually *increase* the effort needed. This is useful if we need to do something which requires accuracy and care, such as holding a glowing piece of coal with tongs or using tweezers to pull out a splinter.

Figure 3.31 shows more examples of levers being used. Explain how these work as machines. Show the effort, the resistance and the fulcrum in each case.

Figure 3.31

ACTIVITY **3.5** **INVESTIGATING LEVERS**

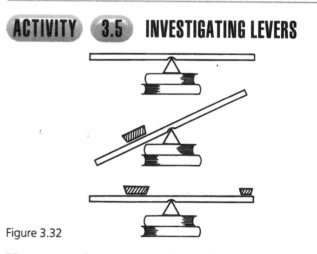

Figure 3.32

You are going to investigate the relation between effort and resistance in a simple lever.

You will need a metre rule, several weights, a wedge and a stack of books or a block of wood.

1 Set up the metre rule on the wedge and adjust its position until it balances.
For the rule to be balanced, balanced forces must be acting on it. What are these forces?

2 Take a weight of, say, 10 N and place it about 30 cm from the fulcrum and to the left of it.
What happens to the rule?

3 Use another weight to balance the rule again.
On which side do you put this weight?

4 Measure the distance of the second weight from the fulcrum.

5 Call the weight on one side, say the left side, the resistance. The distance of this resistance from the fulcrum is called the **resistance arm**. The weight needed for balance on the right side is the effort. The distance between the effort and the fulcrum is called the **effort arm**. Copy the table below.

Trial	Resistance (R)	Effort (E)	Effort arm (A_E)	Resistance arm (A_R)	$E \times A_E$	$R \times A_R$
1						
2						
3						
.						
.						

6 Use various weights as the resistance at different distances (A_R) from the fulcrum. Each time find the effort necessary to balance the rule, and the distance of the effort from the fulcrum (A_E). Record your observations in the table.
In each case, what do you observe about the values of $R \times A_R$ and $E \times A_E$? Your answer to this question should form a consistent pattern. This pattern can be stated as a law, called the law of levers. From your observations, state this law.

Figure 3.33   *A see-saw*

On your own

1 Explain how a see-saw works (see Figure 3.33).

2 Design a lever system to find the weight of someone in your class.

3 How could you use the arrangement with the metre rule, the wedge and weights to find the weight of the metre rule?

Moment of a force

When a force acts on an object which is pivoted, this may cause turning motion, as we have seen. It is possible to define the effect of such a force as the **moment** of the force. The moment depends on both the size of the force and the distance from the pivot of the point of action of the force. We can then say that:

moment of a force = force × perpendicular distance of force from pivot

What will be the unit of a moment?
In the diagram shown below, the force P (300 N) is shown as acting at a distance of 2 m from the pivot O, such as to cause anticlockwise motion. The moment of the force P about the pivot O is then:

P (300 N) × 2 m = 600 N m

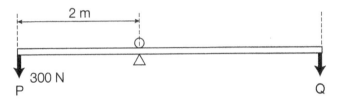

Figure 3.34

If another force Q acts at a distance of 3 m as shown, this would cause clockwise motion. Calculate the size of Q for the lever to be balanced.

The pulley

The **pulley** is another type of machine. In a pulley system the pulley can be fixed or it can move.
When you have a single pulley, the effort is not reduced. Why is the pulley still useful?

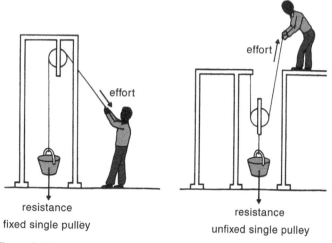
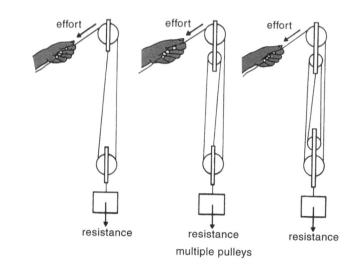

Figure 3.35

Sometimes more than one pulley is used to do work. Look at the diagrams above. In which one will less force be required to lift the object? Explain why.

1. Think of a type of work. Design a machine to do this work. Remember that, in science, work is only done when a force moves something through a distance.

2. Design a machine that will help someone with a physical disability do a particular kind of work.

Figure 3.36 *A submersible has to withstand great pressures*

3.7 Pressure and its measurement

The air which forms the Earth's atmosphere stretches a long way upwards. Although it is light, air has weight – for example, the air in an average room might weigh about 500 N, similar to the weight of a student!

The particles (molecules) in air move about in all directions, and this constant movement is the source of pressure on objects. There are many particles in the air at sea level, so the pressure is quite high.

A solid object exerts a pressure when it is placed on a surface, and pressure is exerted by liquids also. People who explore the oceans have to be very careful because of the very large forces exerted at great depths. The machines used for exploration have to be constructed so that they can withstand these very high pressures.

ACTIVITY **3.6** **DEMONSTRATION OF AIR PRESSURE**

This can be demonstrated by your teacher, who will need a metal can, a vacuum pump, and Magdeburg hemispheres (if available).

1. The can should be connected to the vacuum pump. The pump can then be used to remove the air from the can.

What do you observe as the air is removed?
How can you explain your observation in terms of the differences in pressure **inside** and **outside** the can?

2 The two Magdeburg hemispheres should be fitted together and then connected to the vacuum pump. The pump can then be used to remove the air from the two hemispheres. Can you or any other student separate the hemispheres when the air has been removed? How can you explain this in terms of the differences in pressure **inside** and **outside** the hemispheres?

The pressure of the air is made use of in many different applications. For example, when you suck up liquid through a drinking straw, you suck air into your lungs from the straw. The pressure of the atmosphere pressing down on the surface of the liquid is now greater than the pressure of the air in the straw so the liquid is pushed up through the straw and into your mouth. When you place a moistened rubber sucker on a smooth vertical surface and press hard to remove the air, the sucker will stay in place because the pressure of the atmosphere holds it against the surface.

ACTIVITY 3.7 MEASURING THE PRESSURE OF THE AIR

Your teacher will demonstrate the use of a **mercury barometer** to measure the pressure exerted by the air. S/he will need mercury (which has to be handled very carefully, because it is very poisonous), a thick-walled glass tube, a funnel, retort stand and clamp and a container.

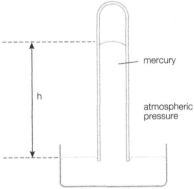

Figure 3.37 *A mercury barometer*

1 The tube should be nearly filled with mercury, using the funnel for pouring the mercury into the tube.

2 The tube should then be inverted *slowly* a number of times with a finger kept over the open end, to collect any air bubbles.

3 The tube should then be filled with mercury and inverted (slowly) into a bowl of mercury – it will be necessary to keep a finger over the end of the tube. The tube can be kept in place using the clamp.

4 What do you observe about the height of the column of mercury in the tube?

5 How can you explain your observation in terms of the pressure exerted by the air?

6 Would you expect that the *vertical* height of the column of mercury would remain the same if the tube is tilted? Would it make any difference if the tube was wider?

3.8 Density and its measurement

We often say that a substance (such as iron or steel) is *heavier* than another material (such as wood or plastic). We can really only compare the substances if we have the same amount (or same mass) in each case – it would be unfair to compare a needle made of steel with a large log of wood!

This means that we have to keep the *volume* constant if we want to compare different substances. In this way we can compare the 'heaviness' of the substances. If the volume is constant in each case, we can then measure the *mass* of each substance. The **mass per unit volume** of a substance is known as its **density**.

It is important to know about the density of solids, liquids and gases. For example, the pressure exerted by a liquid will depend on its density. You can think about this in terms of the mass of a column of liquid which has a cross-sectional area of 1 square centimetre. The greater the density of the liquid, the greater the mass of the column and therefore the greater the pressure exerted by the liquid.

If we know both the mass and the volume of a substance, we can calculate its density. For example, the mass of a piece of lead is 275 g and its volume is 25 cm^3. The density of lead is: $275/25 = 11$ g/cm^3.

ACTIVITY 3.8 MEASURING DENSITY

You will need a glass block, a balance, a ruler, a pebble, a measuring cylinder and a beaker.

1 a Find the mass of the glass block using the balance.

b Now measure the lengths of three different sides of the block.

c How will you calculate the volume of the block?

d Given that density (*d*) is mass/volume, what is the density of the glass block?

2 a Find the mass of the pebble.

b Measure the volume of the pebble by measuring how much water is displaced by the pebble in a measuring cylinder.

c Calculate the density of the pebble.

3 a Find the mass of the beaker.

b Measure out a known volume (perhaps 50 cm^3) of water in a measuring cylinder.

c Transfer the water from the measuring cylinder to the beaker.

d Now find the mass of the beaker plus the known volume of water.

e You know the volume of the water. Can you calculate from your measurements the mass of the water? What is the density of water?

Summary

These are some of the main ideas that you have learnt in this unit:

- A force is a push or a pull.

- A spring balance can be used to measure forces.

- The unit of force is the newton.

- The force of gravity gives objects their weight.

- Friction is a force which opposes motion. Friction occurs when two surfaces move over each other.

- Velocity describes the speed and direction of movement of an object.

- An object is in uniform motion or constant velocity, while its speed and direction of movement remain unchanged.

- An object remains in uniform motion unless an unbalanced force acts on it.

- An increase in speed is called acceleration.

- For every force, there is an equal force in the opposite direction.

- Work is done when a force causes something to move.

- Work is measured in newton metres.

- Machines are used to make work easier.

- When machines are used in practice, work input is always greater than work output.

- The moment of a force is defined as the force multiplied by the perpendicular distance of the force from the pivot.

- Pressure is defined as force per unit area, and is measured in units of newtons per square metre and pascals.

- Barometers can be used to measure pressure.

- Density is defined as mass per unit volume, and is measured in units of grams per cubic centimetre.

QUESTIONS

1 If a constant force is applied to an object which can move freely, the effect of the constant application of the force *may* be to change any of the following *except*

A velocity
B acceleration
C momentum
D kinetic energy

2 Which of the following is *not* an example of the law of action and reaction?

A The removal of water from clothes in a dryer
B The recoil of a gun when fired
C The difficulty of stepping out of a small boat
D The breaking of a cup when it is dropped on the floor

3 A mass of 2 kg is placed on a horizontal surface. The area in contact with the surface is 50 cm². The pressure exerted on the surface is

A   40 kg/cm²
B   0.04 kg/cm²
C   100 kg/cm²
D   10 kg/cm²

4 Machines are used to make work easier. In which of the following implements/machines do you *not* take advantage of a large moment provided by a relatively small force?

A   A claw hammer for taking out nails
B   A pair of wire cutters
C   The foot treadle on a sewing machine
D   A pair of scissors

5 Which of the following diagrams shows the arrangement of forces which will give block M the *greatest* acceleration?

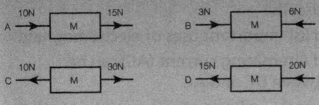

Figure 3.38

6 A metal cube, weight 100 grams, of side 2 cm is lowered into water in a measuring cylinder. The volume of water appears to increase by 8 cm³. This shows that

A   a solid displaces its own weight of water
B   a solid displaces its own volume of water
C   the measuring cylinder is calibrated directly in cm
D   the density of water is 1 g/cm³

7 Three glass tubes (thick walled) are set up as shown in the diagram below. The lengths of the tubes are as follows:

A – 65 cm
B – 95 cm
C – 110 cm

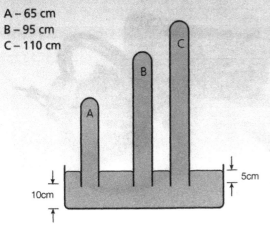

Figure 3.39

The tubes are closed at one end and each tube is completely filled with mercury. The tubes are inverted without allowing any mercury to escape, and placed with the open end under a pool of mercury as shown. The pressure of the atmosphere is known to be 76 cm Hg.

i) Copy the diagram, and show clearly what happens to the mercury in each tube.
Make sure that all distances are marked clearly on your diagram.

ii) Tube B is *pushed down* by a further 5 cm into the bowl/container. Describe briefly what happens to the mercury in the tube.

iii) Describe what will happen if Tube C is *raised* by 7.5 cm.

8 A metre rule is marked off at 10 cm intervals. The metre rule is placed on a triangular fulcrum and a number of identical masses placed at various points on the rule. The diagrams X, Y and Z show three different arrangements of the masses on the rule. In *each* case, describe clearly, giving your reasons, what will happen to the rule. For example, will it be balanced, will it turn clockwise, etc.?

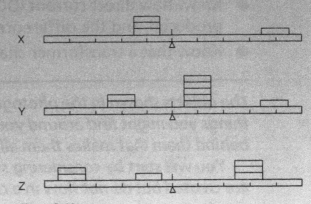

Figure 3.40

9 When a woman walks across a wooden floor in shoes with very high heels it is possible that she may cause more damage to the floor than an animal with a much larger weight such as an elephant. Explain why this might be the case.

10 a If the density of a piece of wood is 0.6 g/cm³ what is the mass of

(i) 1 cm³   (ii) 5 cm³   (iii) 10 cm³?

b What is the density of a substance of

(i) mass 200 g and volume 25 cm³
(ii) volume 9 m³ and mass 27 kg?

c The density of the metal gold is about 19 g/cm³. Find the volume of

(i) 76 g
(ii) 190 g of the metal.

UNIT 4

Electrical and magnetic energy

OBJECTIVES

- Carry out simple experiments to show that electric charges can be created by rubbing suitable substances
- Know that the flow of charge in a circuit is the current, measured in amperes
- Understand and be able to carry out calculations involving Ohm's Law
- List the uses of fuses in electrical circuits
- Make a simple electromagnet and list important uses of electromagnets
- Know how direct current (DC) and alternating current (AC) can be produced, and the differences between DC and AC
- Know that a transformer changes an AC voltage at one value to another

The objects shown in the photographs will be familiar to you as things you might find around your home. But what is the force behind them that makes them all work? It is electricity.

You will start by considering static electricity, in which charges are created but do not flow in a circuit. You will then use electric circuits to measure and calculate current, voltage and resistance.

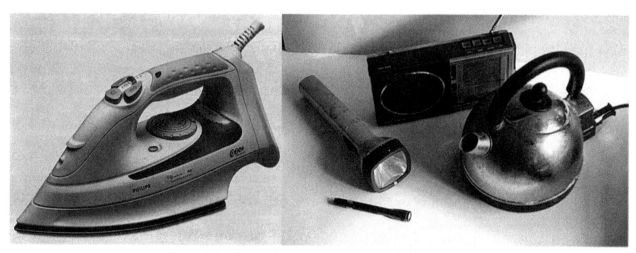

Figure 4.1 *All these household appliances work by electricity*

4.1 Static electricity

In a dry climate, a nylon garment will often crackle when it is taken off. We can explain this by saying that the garment has become charged with static electricity. The sounds heard are caused by electrical sparks which can be seen in the dark. It is possible to create electrical charges by rubbing suitable items with a cloth.

 ACTIVITY **4.1** **POSITIVE AND NEGATIVE CHARGES**

You will need two strips of polythene, a piece of another plastic (e.g. cellulose acetate), two paper holders and some thread, and a piece of dry cloth.

1 Rub one of the pieces of polythene with the cloth. Hang it up as shown in the diagram.

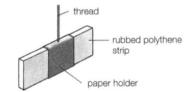

thread
rubbed polythene strip
paper holder

Figure 4.2

2 Now rub the second piece of polythene, and bring it close to the suspended piece. What do you observe?

3 Rub the other piece of plastic and bring it close to the suspended piece of polythene. What do you observe?

4 Can you explain your observations?

You should be able to show that there are two different types of charge, known as **positive** and **negative**, and that like charges repel and unlike charges attract.

We can explain the existence of two different types of charge by looking at the structure of the atom. Atoms are thought to contain two types of charged particle, **protons** (positive) and **electrons** (negative). The protons are found in the nucleus of the atom, while the electrons move around the nucleus. When some items are rubbed, electrons can be transferred from one material to another. For example, when cellulose acetate is rubbed, electrons move from the acetate to the cloth, leaving the plastic with fewer electrons than at the

Figure 4.3 *A good example of the effect of static electricity is lightning*

beginning. Since the protons have not moved from any of the atoms, the plastic becomes positively charged – the cloth is negatively charged.

4.2 Electrical circuits

We looked at electrical circuits in Book 1, Unit 8 (8.9–8.11). Look at the circuit diagrams below and try to remember what the various types of circuit are called. Copy one of them into your note book and label the source of electricity, the bulb and the switch.

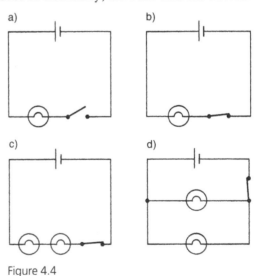

Figure 4.4

Which bulbs, if any, will light in the circuits shown in Figure 4.5?

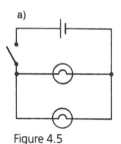
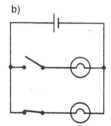

Figure 4.5

Terminals of an electric cell

As you have seen, we use the symbol —|⊢— to represent an electric cell. The long line represents the **positive terminal** of the cell, and the short line represents the **negative terminal**. The photograph below shows the negative and positive terminals on a typical electric cell. You need to be clear which terminal is which when you use the instrument you are going to learn about in the next section.

Figure 4.6

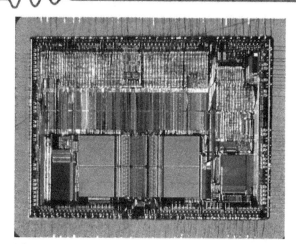

Figure 4.7

Many electronic instruments contain complex circuits consisting of many components packed together in a very small space. These circuits are called 'integrated circuits', or ICs.

4.3 Current

When we make a circuit with a cell, a switch and a bulb, and the bulb lights we say that electricity **flows** in the circuit. This flow of electricity is a flow of electric **charge**. The rate of flow of this charge is called the **current**. The current is measured in units called **amperes** or **amps**. The symbol for the ampere is A.

We can measure the current with an instrument called an **ammeter**. The symbol for an ammeter in a circuit diagram is —Ⓐ— .

Figure 4.8   *Ammeters*

ACTIVITY 4.2 USING AN AMMETER

You are going to use an ammeter to measure the current flowing in circuits.

You will need an electric cell in a circuit board, some wire, crocodile clips, two bulbs and an ammeter.

❶ Make a circuit like the one shown below.

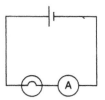

Figure 4.9

Make sure the ammeter is connected correctly. One terminal of the ammeter should have a plus sign by it. It may also be coloured red. This should be connected to the positive terminal of the cell. It need not

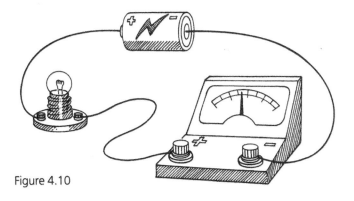

Figure 4.10

be *directly* connected to the positive terminal: there may be other circuit components between it and the terminal. But it should be connected to the wire that is eventually connected to the positive terminal.

The other terminal of the ammeter should have a minus sign by it. It may also be coloured black. This should be connected to the wire that is eventually connected to the negative terminal of the cell. When you have set up your circuit, it should look like the one shown in Figure 4.10.

What is the reading on the ammeter?

2 Now set up the circuit shown below.

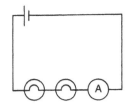

Figure 4.11

What is the reading on the ammeter?
In which circuit did the ammeter give the higher reading?

Later in this unit, you will see why the current was different in these two circuits.

It is important to make sure that you use an ammeter with the right **range** to measure the current in the circuit. Some ammeters have one fixed range, e.g. 0–5 A. Before you use such an ammeter, you should know whether the current in the circuit is likely to be greater than 5 A – if so, you should use another ammeter, with a greater range.

Other ammeters have different ranges. For example, you might have an ammeter with three different possible ranges:

0–50 mA, 0–500 mA, 0–5 A.

When using such an ammeter, you need to select the range which will enable you to obtain the most accurate measurement of the current in the circuit. For example, why will the range 0–50 mA give a more accurate reading of a current of 10 mA than the range of 0–5 A?

If you wanted to find what current was flowing in a circuit which contained three bulbs in series, which of the following arrangements would you use?

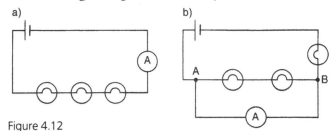

Figure 4.12

You should use the first arrangement because the current which flows through the three bulbs then flows through the ammeter. In the second arrangement, the current would split into two. Part of it would go from A to B through the two bulbs. The other part would go from A to B through the ammeter. So the ammeter would not be measuring all the current in the circuit, but only that part which was not going through the first two bulbs.

On your own

Set up the circuits shown in the three diagrams. What reading do you get on the ammeter in each case? What can you infer about the current flowing through a series circuit?

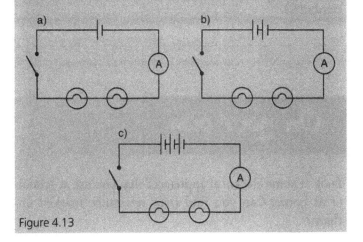

Figure 4.13

The coulomb – a unit of charge

We have said that the current in a circuit may be defined as the rate of flow of electrical charge around the circuit. The current is measured in amperes (A). If the current in a circuit is 2 amperes, this is represented as 2 A.

The unit of charge is the coulomb, which is defined with reference to the unit of current, the ampere:

One coulomb (1 C) is the charge passing any point in a circuit when a current of 1 A flows for one second.

For example, a charge of 5 C would pass each point in a circuit in one second if the current was 5 A. In two seconds, 10 C would pass each point, and so on.

The general symbol for electrical charge is Q. We can express the relationship between charge (Q), steady current (I) and time (t) as follows:

$Q = It.$

Resistance

Conductors allow electricity to flow through them. Metals are good conductors of electricity. However, not all metals are equally good conductors. Some metals allow electricity to flow through them more easily than other metals do.

How easy it is for electricity to flow through a conductor depends on a property of the conductor called the **resistance**. If the resistance is very high, the current will be low. If the resistance is very small, the current will be high.

Resistance is measured in units called **ohms** – symbol Ω.

On your own

Look at some electrical appliances that you use at school or at home. Can you find their resistance marked on them?

You are going to design an experiment to investigate whether the length and thickness of a wire affect its resistance.

You will need copper wire connected to an electric cell, an ammeter, crocodile clips, high resistance wire (e.g. nichrome) of varying thicknesses.

1 Decide how you will set up a circuit to measure any change in resistance in different samples of nichrome wire.

2 Draw a diagram of the circuit you will use.

3 Decide how you will test whether the length of the nichrome wire affects its resistance.
What will you keep the same in this test?

4 Carry out the test you have decided on and record your results.
Do you think that the length of a piece of wire affects its resistance?

5 Repeat the test on a number of lengths of wire to see if your results support your answer in step 4.

6 Now repeat steps 3 to 5, but this time test whether the thickness of the nichrome wire affects its resistance.

Resistance of a wire

Material

If you take two pieces of wire of the same length and same thickness, but made of different materials, the resistances will be different. For example, platinum wire has a higher resistance than copper wire. Copper has a very low resistance, which is why we use it in electric circuits.

Length

Imagine you have two pieces of wire made from the same material and of the same thickness, but of different lengths. Which do you think will have the greater resistance?

The longer the wire, the bigger is the resistance. The current has to 'force its way along' a longer length of wire.

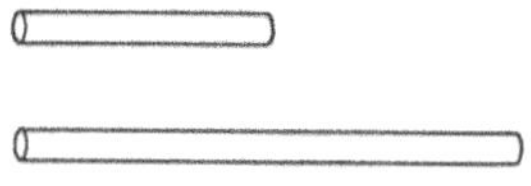

Figure 4.14

Thickness

Imagine you take two pieces of wire made of the same material and of the same length, but of different thickness. Which do you think will have the greater resistance?

Figure 4.15

The thinner wire has the greater resistance. The current has to 'squeeze its way through' a smaller space.

You should now be able to understand your result in step 2 of Activity 4.2. You should have found that the current was higher when you had one bulb in a circuit than when you had two in series. Electric light bulbs are made with very thin lengths of wire. Look at a bulb to see this. When you have two bulbs in series you have twice the length of this thin wire in the circuit. This means that the resistance is higher than with one bulb. Since the resistance is higher, the current is lower, as you should have found out from the activity.

Resistors

Sometimes we need to protect the parts of a circuit from becoming damaged by too high a current. We put a resistance of a certain value into the circuit. This resistance is called a **resistor**. A resistor helps to control the size of a current in a circuit.

The resistance of some resistors can be adjusted. Such resistors are called **variable resistors** or **rheostats**. The diagrams below show a rheostat in a circuit. By sliding the contact along the length of the resistor, the value of the resistance changes.

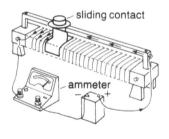

sliding contact

Current flows through *short* length of resistance wire. The current is large

ammeter

sliding contact

Current flows through *long* length of resistance wire. The current is small

ammeter

Figure 4.16

Figure 4.17 *Rheostats are used in many electrical devices such as a lighting dimmer switch and a volume control*

4.5 Fuses

Electrical devices which go wrong are very dangerous. They may give someone who touches them an electric shock. Very often the fact that a device is not working properly may cause a large current to flow through it. This adds another danger: the wires in the device become hot. They may melt and start a fire.

For these reasons we need **fuses** to protect circuits when the current becomes too large. A fuse contains a very thin wire. The fuse is placed in series in the circuit. If the current gets higher than it should be, the wire gets hot and melts: we say the fuse 'blows'. This breaks the circuit, so that the electricity stops flowing and the circuit is protected from further damage.

Figure 4.18 *Fuses of different value*

Each fuse has the value of the current it can take marked on it (see Figure 4.18). Different values of current will be suitable for different electrical devices.

When a fuse blows, the device cannot be used again until it is mended and the fuse replaced. If a fuse blows, do not replace it until you have found out what the original fault was and make sure that this fault has been put right.

Figure 4.19 *Industrial circuit breakers (top). A domestic circuit breaker (bottom) protects the user*

Many houses use circuit breakers instead of fuses. They work in a similar way to fuses.

4.6 Electrical energy

You know that electricity is a form of energy. How can other kinds of energy be converted into electrical energy?

We can get electrical energy from:

Chemical energy: When chemical reactions take place in an electric cell, electrical energy is produced.

Light energy: Spacecraft and satellites sometimes use cells called 'photocells' which convert light energy to electrical energy. Some pocket calculators have panels which convert light energy.

Heat energy: When two conductors, for example a copper wire and an iron wire, are joined together at the

a

b

c

Figure 4.20 *Examples of sources of electrical energy obtained from other forms of energy*

ends and the ends are kept at different temperatures, electrical energy is produced.

Heat energy obtained from inside the earth can also be converted to electrical energy.

Kinetic energy: When running water, steam or wind are used to turn generators, kinetic energy is converted into electrical energy.

Nuclear energy: Energy stored in the nuclei of certain atoms is converted into electrical energy in nuclear power stations.

Electricity – electrons in motion

A scientist called J. J. Thomson did many experiments and discovered what causes electricity. He explained that electricity is obtained when electrons move in the circuit. You have already learnt that electrons are one of the particles in an atom. You may remember that each electron carries an **electric charge**.

Electrons can flow through an electric circuit, and the speed at which they flow through the circuit can vary. The **rate of flow of electrons** through a circuit is called the **electric current**. The symbol often used for electric current is the letter *I*.

> Did you know that a very small current of about 0.005 A can be harmful to the body?

Electric voltage

In order to get the electrons to move from the source of the electricity, they must be given energy in circuits. Electric cells are often the source of the energy. The energy given to electrons is called the **voltage**. Voltage is measured in units called volts (V). The volt is named after the Italian scientist Alessandro Volta who did some very early work in electricity and invented the electric cell.

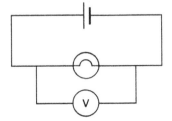
On your own

Look at an electric cell. It should have its voltage marked on it. What is the voltage of the cell?

When the electrons flow through the components in a circuit they give up some of their energy. We say that a voltage drop takes place. The drop between two points in a circuit is called the **potential difference**. Potential difference is measured in volts.

The instrument used to measure voltage and potential difference is called a **voltmeter**.

Figure 4.21

Look at the circuit in Figure 4.21. The voltmeter is being used to measure the potential difference across the bulb. Why is it placed in parallel with the bulb?

Voltage, current and resistance

Ohm did many experiments to find out what happens to the current when the voltage is changed or when the resistance is changed. He found a pattern in his results.

When the voltage was increased, the current increased in the same proportion. For example, if the voltage was doubled, the current doubled. When the voltage was decreased the current decreased in the same proportion. When a pattern like this is observed, we say that the two variables (voltage and current) are **directly proportional** to each other.

Ohm was able to show how voltage, resistance and current are related to each other. We state the pattern his results showed as a law.

> **Ohm's Law: The current in a circuit is directly proportional to the voltage and inversely proportional to the resistance.**

We can state Ohm's Law as an equation:

current (I) = voltage (V)/resistance (R) : $I = V/R$

which is the same as saying $V = IR$
Using this equation we can calculate the value of one variable if we know the values of the other two variables.

Example
What is the value of the resistance in a circuit if the voltage is 20 V and the current is 4 A?

Steps

a   Write down what you are given. ($V = 20\,V, I = 4\,A$)
b   What do you need to know? (R)
c   What is the relationship? ($V = IR$)
d   Substitute values in the equation. ($20 = 4 \times R$)
e   Calculate the unknown value. ($R = 20/4 = 5\,\Omega$)
f   State the answer. (Resistance = $5\,\Omega$)

> Did you know that the resistance of the human body decreases from about 500 000 Ω when the skin is dry to about 100 Ω when the skin is wet? You should make sure your hands are dry when you touch any electrical instrument. Why?

ACTIVITY 4.4 MORE THAN ONE RESISTOR

You are going to investigate the total resistance when more than one resistor is used.

You will need two resistors of known resistance (e.g. of 100 Ω and 50 Ω), dry cells, wires, crocodile clips, a voltmeter and an ammeter.

❶ Design and carry out an experiment to find the resistance of the two resistors when they are in series in a circuit. Draw a circuit diagram.

❷ Design and carry out an experiment to find the resistance of the two resistors when they are in parallel in a circuit. Draw a circuit diagram.

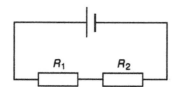

Figure 4.22

The two resistors above are connected in series. The total resistance in this circuit is found by adding the resistance values:

total resistance = resistance of R_1 + resistance of R_2
$R_{total} = R_1 + R_2$

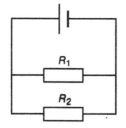

Figure 4.23

The two resistors above are in parallel. The total resistance in this circuit is found using the equation:

$1/R_{total} = 1/R_1 + 1/R_2$

Suppose R_1 and R_2 both had a value of 10 Ω. Then:

$1/R_{total} = 1/10 + 1/10 = 2/10$
$R_{total} = 10/2 = 5$ Ω

The total resistance is 5 Ω.

Electric power

When we use electricity in the home, we have to pay for the energy we use. This energy is the energy given up by electrons when a current flows through a circuit. For example, when a current flows through the filament of a light bulb, the electrons give up energy which is converted into light and heat energy. When current flows through an electric cooker, the electrons give up energy which is converted into heat energy.

How can we know how much energy each device will use? Look at a light bulb. What is written on the glass? When you read the rating of a bulb for example you will see values like 40 watts or 40 W, 60 W and so on. These values tell how much electrical energy is used by the bulbs in a given time. They are values of **electric power** or the **power rating** of bulbs. Power is the rate of using energy. That is:

power = energy/time

The unit of power is called the watt, symbol W, named after the Scottish scientist James Watt, see page 84.

Do you remember that energy is measured in units called joules? One watt means the same as one joule per second. So, if a light bulb is marked '60 W', this means that it uses up 60 joules of energy in each second.

In the sections on electric voltage and electric current you should remember that the voltage is a measure of the energy given to the electrons as they move in the circuit. The current is the rate of flow of the electrons in the circuit. We can calculate the power in a circuit using the equation:

power = voltage × current

Example
A television is connected to a 120 V outlet and draws a current of 1.2 A. What is its power rating?

power = voltage × current
= 120 × 1.2 W
= 144 W

The television has a power rating of 144 watts.

1 What is the power rating of an electric can opener used in a 110 V circuit and drawing 2.2 A?

2 What is the current used by a 1200 W electric kettle when it is connected to a 110 V outlet?

Selling electricity

How do we pay for our electricity? We pay for the number of units of electricity we use over a given time. For example, we may have to pay for the units used in a month. The units used in calculating electricity bills are kilowatt-hours, symbol kWh.

1 kilowatt = 1000 watts

If we use one kilowatt-hour of electricity, how many joules of electrical energy have we used up?

Look at these meters:

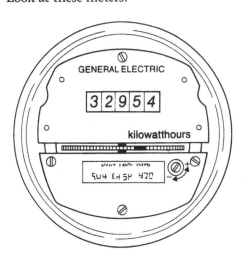

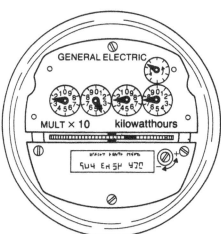

Figure 4.24

It is easy to read the first one. To read the meter with the dials, follow these steps:

a   Read the dials from left to right.
b   Multiply the figure you get by 10.

You should get 24 270.

This is the number of kilowatt-hours that have been used since the meter started.

Look at these diagrams:

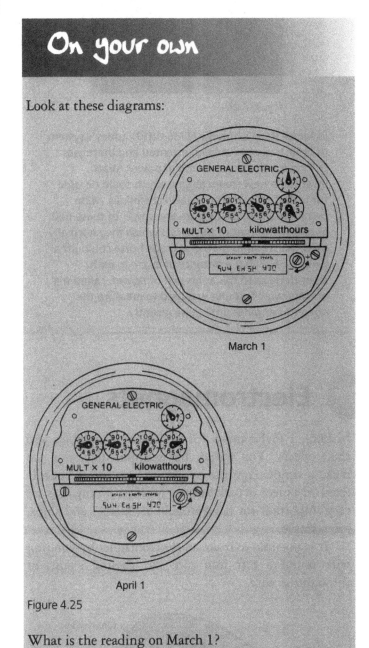

March 1

April 1

Figure 4.25

What is the reading on March 1?

What is the reading one month later on April 1?

How many kilowatt-hours of electricity have been used?

If the cost of electricity is 17 cents for one kilowatt-hour, what will the bill be for the month of March?

Figure 4.26

In James Watt's lifetime (1736–1819), today's system of units had not yet been invented and there was no commonly accepted unit of power. Watt invented a new steam engine which could be used for lifting containers of coal in coalmines. Mine owners wanted to compare the power of these new engines with that of the horses which they normally used. Watt found that an average horse could lift a weight of 2500 N a distance of 30 cm in each second. He used this as his unit of power, called the horsepower. This unit was used to measure the power of engines until quite recently.

4.7 Electromagnets

In Book 1 (8.12 and 8.13) we looked at magnets and their effects. You found, for example, that you can make a simple compass by using a magnetized needle fixed on a piece of cork floating in water. Compasses are very important for navigators in aeroplanes and ships, for example.

It is possible to make an electromagnet by winding wire round a soft iron core and passing a current through the wire.

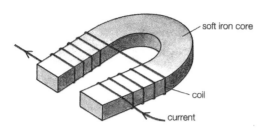

Figure 4.27   *Making an electromagnet*

The magnetism of an electromagnet is temporary, and depends on the current flowing through the wire and its effect on the soft iron core. If there is no current flowing there is no magnetism.

The strength of the magnetism (magnetic field) can be changed by altering:

a   the current in the wire
b   the number of turns of the wire coil
c   the distance between the poles of the core.

Figure 4.28a   *Electromagnets are used in telephones*

Figure 4.28b   *Very strong electromagnets are used in industry to lift metal objects*

Electromagnets are also used in many household devices, such as the door bell, and telephone as described on page 85.

ACTIVITY 4.5   MAKING AN ELECTROMAGNET

You are going to make a magnet using the electricity in a circuit.

You will need a bar of soft iron (or a large iron nail), two electric cells, a switch, crocodile clips, a long piece of wire and some paper clips.

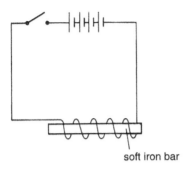

Figure 4.29

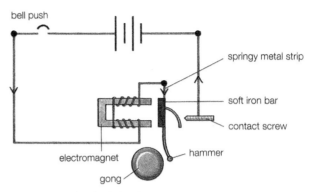

Figure 4.30   *A doorbell circuit*

1. Set up a circuit as shown in Figure 4.29. Make sure that the wire is long enough for the iron bar to be moved about easily. Put about five turns of wire round the iron bar.

2. Close the switch and hold the iron bar over some paper clips.

   What do you observe?

3. Increase the number of turns of wire around the iron bar. Close the switch and hold the bar over some paper clips again.

   Did it pick up more or fewer clips?
   In which circuit did you get a stronger magnet?
   How do you know?

4. Put another cell into the circuit and keep the same number of turns of wire around the iron bar as you had in step 1. Close the switch and hold the bar over the paper clips.

   What do you observe?
   How many clips did the bar pick up this time?
   Has the extra cell increased the strength of the magnet?

5. Remove the iron bar from the circuit.
   Now hold it above the paper clips.
   Does it pick any up?

Doorbell

When someone pushes the bell push, the circuit is completed, and current flows through the electromagnet. Once the current flows through the electromagnet it becomes magnetized, and therefore attracts the soft iron bar (also known as the armature). This causes the hammer to hit the bar – but the circuit has now been broken at the point of the contact screw, because the iron bar has moved.

If the circuit is broken, there is no current flowing in the coil round the electromagnet, so it loses its magnetism. This in turn means that the iron bar is no longer attracted to the electromagnet, so the metal strip (which is springy) pulls the iron bar back. This completes the circuit again, so current flows again. This sequence continues as long as the person keeps pushing the bell push.

Telephone

The receiver of a telephone contains an electromagnet. The microphone in the speaking end of a telephone converts sound energy into electrical energy (see Figure 4.28a). The electrical energy travels from the speaking end of one telephone to the receiving end of another. It is changed into sound by the receiver.

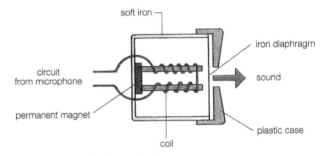

Figure 4.31   *A telephone circuit*

The current received from the speaking end (microphone) passes through the coils of the electromagnet. This current varies depending on the amount of sound energy received from the microphone at any given moment. The diaphragm (iron) is pulled towards the electromagnet. The distance which the diaphragm is pulled depends on the current. The diaphragm therefore moves in and out and produces sound waves. These waves are a copy of those received from the microphone.

Electric current: direct current (DC) and alternating current (AC)

It might be helpful for you to look back at Sections 8.7–8.10 in Book 1 before you work through this section.

So far, we have made use of electrical circuits in which the current has been generated by a dry cell, similar to those used in transistor radios, for example. Earlier in this unit, we defined electric current as the flow of charge around a circuit. There are a number of different devices that produce electrical energy from chemical energy. These include the dry cell and the wet cell – a common example of the latter is the lead-acid accumulator, still used in many cars and referred to as the 'battery'.

The dry cell

Dry cells are used in devices such as flashlights, transistor radios, electric razors, etc. The simplest dry cell draws on the chemical energy arising from the reaction between zinc (Zn) and ammonium chloride. In this reaction, hydrogen gas is produced, and collects at the carbon rod which is part of the cell. The gas is removed by the substance manganese(IV) oxide, which is called a 'depolarizer'.

There are now many other types of dry cell, often used in devices such as personal organizers, cameras, etc. The chemical reactions in these cells are different from those in the simple dry cell. Care has to be taken when disposing of dry cells, since the chemicals may be toxic.

Dry cells are used when a steady current is required over a period of time. Once the chemical reaction is complete the cell is dead, and no more current can be obtained from it. It cannot be used again. The dry cell is an example of a primary cell.

The wet cell (lead-acid accumulator)

The lead-acid accumulator is an example of a 'secondary' cell. A secondary cell can be recharged by passing a current through it, in a direction opposite to that provided by the cell.

The lead-acid cell consists of two plates. The positive plate consists of lead(IV) oxide and the negative plate is lead. Sulphuric acid is the conducting liquid. When the cell supplies current, both plates change to lead sulphate, and the sulphuric acid gradually becomes more dilute as the sulphate is used up.

When the relative density of the acid reaches a particular level (about 1.18) the cell is said to be discharged and no more current can be obtained from it. The cell can then be recharged by supplying a current to it, with the positive of the supply being connected to the positive plate in the cell. The current supplied can be adjusted and the cell charged for the correct number of hours.

The 'battery' in many cars consists of six lead-acid cells, connected in series, with an output voltage of 12 V. These 'batteries' are very heavy because of the density of lead, and many attempts have been made to find lighter materials. Cars powered by electrical cells are becoming more common in some parts of the world, with the state of California in the USA leading the way in development, largely because of state legislation about pollution.

Both the dry cells and wet cells supply what is called **direct current** (DC), in which the direction of flow of current remains the same.

Alternating current

The electricity supplied to our homes (see Book 1, Section 8.7) is not direct current. Electricity for household use is supplied through some form of electricity generation. This may involve the burning of a fuel, such as coal or gas, so that chemical energy is ultimately converted to electrical energy. In some countries, such as France and Japan, a high proportion of the electricity generated makes use of nuclear energy. In other countries, such as Canada, hydroelectricity is an important source of power.

The electricity supplied to our homes is alternating current, in which the direction of current changes rapidly. For example, if it changes direction 50 times a second, the frequency of the current is said to be 50 c/s. You may have to take care when purchasing some electrical appliances, since the frequency can vary from country to country, some using 50 c/s and some 60 c/s.

In the laboratory we can demonstrate the generation of alternating current, using a simple apparatus. This is shown in Figure 4.32.

A simple generator requires a rectangular coil, which is placed between the poles of a magnet, as shown. The ends of the coil are joined to two slip rings, on an axle. Carbon brushes press against the slip rings. When the

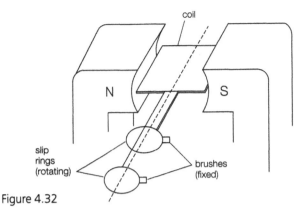

Figure 4.32

coil rotates, it cuts through the magnetic field and induces an electromotive force.

The transformer

A transformer is used to change an alternating current voltage (potential difference) from one value to another. If the change is an increase, the transformer is a step up transformer. If the change is a decrease, it is a step down. In a step up transformer there are *more* turns in the *secondary*; whereas for a step down transformer there are *fewer* turns in the *secondary*. An example of a simple transformer is shown in the diagram.

Transformers are used in supply systems in which the electricity is first generated and then distributed at a very high voltage – perhaps as high as 25 000 V or more. In some large national systems, the electricity is first generated at about 25 000 V and then *stepped up* to 275 000 or even 400 000 V and distributed through the system. It is then *stepped down* at smaller sub-stations to the locally used voltage – for example 220 V or 110 V – and distributed to homes and businesses. See if you can find out how electricity is generated and then distributed in your country.

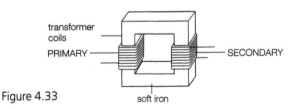

Figure 4.33

Summary

Here is a summary of the main ideas you have learnt in this unit:

- Electric charges can be created by rubbing objects with cloth, for example.

- There are two types of electric charge, positive and negative.

- When current flows in an electric circuit, this is a flow of electric charge.

- The current in a circuit is the rate of flow of electric charge, and the unit of current is the ampere (A).

- An ammeter is an instrument used for measuring current.

- Resistance is the measure of how difficult it is for a current to flow through part of a circuit. The unit of resistance is the ohm (Ω).

- The longer a piece of wire, the greater the resistance; and the thinner the piece of wire, the greater the resistance.

- A fuse is used to stop electricity from flowing through a circuit if the current is too high.

- The energy given to the current (flow of charge) in a circuit is defined as the voltage; the unit of this energy is the volt (V).

- The relationship between the voltage, current and resistance in a circuit is found in Ohm's Law:

    current = voltage/resistance, or $V = IR$.

- Power is the rate of using energy, and the unit is the watt (W).

- In an electric circuit, the power can be calculated by multiplying current and voltage.

- An electromagnet consists of wire wound on a soft iron core: the magnetism is temporary.

- Electromagnets have a wide range of applications.

- The main difference between direct current and alternating current is that the direction of current flow remains the same for direct current, but changes regularly in alternating current.

- Dry cells and wet cells may be used as sources of direct current.

- Transformers are used for changing the value of the voltage of an alternating current supply from one value to another.

QUESTIONS

1 There are two kinds of electric charge. This can be shown by observing that

   A   two charged plastic rods repel each other
   B   a charged plastic rod attracts a charged acetate rod
   C   no substance repels both charged rods
   D   there is a flow of charge around a circuit.

2 The electrical resistance of a coil of wire least depends on the

   A   length of the wire
   B   diameter of the wire
   C   temperature of the air
   D   material of the wire.

3

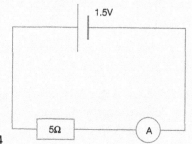

Figure 4.34

Assume that the resistance of the wire used to complete the circuit can be ignored. Use Ohm's Law to calculate the current in the circuit as measured by the ammeter A.

   A   0.3 A
   B   3.0 A
   C   3.3 A
   D   6.5 A

4 A fuse in a domestic electrical circuit

   A   has a very high resistance
   B   always stops electric current flowing
   C   melts when the current through it becomes too high
   D   reacts with substances in the air.

5 Electromagnets *cannot* be used

   A   for lifting heavy metal objects
   B   in telephones
   C   in electric motors
   D   as heating coils.

6 In Figure 4.35, the four lamps are similar. When the current switch is closed, the lamps light. The current flowing through the circuit, as shown by the ammeter A, is 2 A.

   (i) What part of the circuit is shown as F?
   (ii) Calculate the voltage across any one of the lamps.
   (iii) Will the voltage across each lamp be the same? Explain your answer.

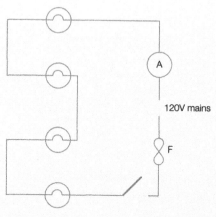

Figure 4.35

   (iv) What would be the power rating (in W) of each lamp?
   (v) What do you think would be the minimum rating for F?

7

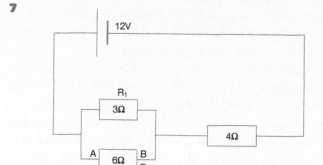

Figure 4.36

In the circuit shown in the diagram above, the potential difference is 12 V.

   (i) Calculate the combined resistance of R_1 and R_2 (3 Ω and 6 Ω).
   (ii) Calculate the current in the 4 Ω resistor.
   (iii) Calculate the voltage across the resistors arranged in parallel – between A and B.
   (iv) Calculate the current flowing through the 3 Ω resistor.

8 The amount of electrical power used in a month (from June 1 to July 1) can be measured by taking the appropriate readings on the meter. The reading on the meter on June 1 is 10 453. The reading on July 1 is 10 896.

   (i) What is the units for the readings?
   (ii) What is the amount of electrical power used in the month?
   (iii) If the cost per unit of power supplied is 10 c, what will be the electricity bill for the month?

Heat energy

UNIT 5

OBJECTIVES

- Distinguish between heat and temperature and use thermometers
- Understand that heat may be transferred in different ways and define, with examples, three different methods of heat transfer – conduction, convection and radiation
- Distinguish between conductors and insulators
- List examples of conductors and insulators
- Perform experiments to show the effect of heat on a solid, a liquid and a gas
- List examples of the effects of expansion – advantages and disadvantages
- Understand the concept of latent heat
- Understand and use the idea of specific heat capacity

We know that the sun is hot. But have you ever thought about how heat reaches us, and how it can be measured?

You will use thermometers to measure temperatures of different substances. You will investigate heat transfer. Some substances are good conductors of heat, others are not. When substances melt or boil, latent heat is involved. Different substances need different amounts of heat energy to raise the temperature of equal masses of the substances.

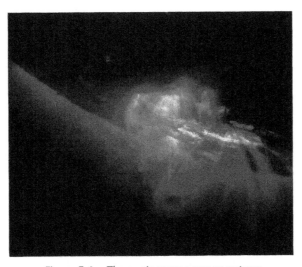

Figure 5.1   *The sun's corona spews out gas*

5.1 Heat and temperature

Heat is not the same thing as temperature. Activity 5.1 will help you to understand one of the distinctions between them. You have used thermometers before – look back at Book 1, Unit 3.8.

ACTIVITY 5.1 RAISING THE TEMPERATURE OF WATER

You are going to see how temperature rises in different volumes of water as they are heated.

You will need water, two beakers or other containers for heating water (they should be identical), two identical sources of heat, two thermometers.

❶ Put a lot of water into one beaker and a little into the other. Record the temperature of the water in each beaker.

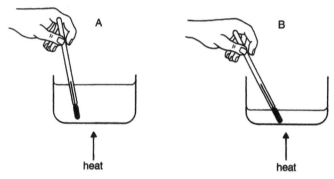

Figure 5.2

❷ Put the beakers over the sources of heat and start heating. (Use the same setting for the heaters in each case; for example, if Bunsen burners are used, they should have flames of the same size.) After one minute record the temperature of water in each beaker.

❸ Record the temperature of water in each beaker at one-minute intervals until the water boils. Show your results on a graph as in Figure 5.3.

Now answer these questions:

a In which beaker did the temperature rise more quickly?

b Was this beaker receiving more heat than the other one?

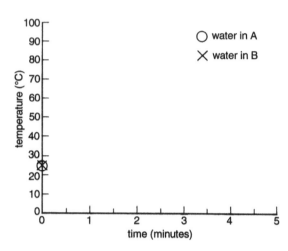

Figure 5.3

c In which beaker did the water take longer to boil?

d Which beaker of water took in more heat energy before the water boiled?

You transferred heat to the water in the two beakers at about the same rate, but the temperature in the smaller volume of water rose more quickly than the temperature in the bigger volume. This shows you that, in at least one way, heat and temperature are not the same thing.

The larger volume of water took longer to boil. You needed to put more heat energy into this beaker of water to make it boil (in other words to raise its temperature to 100°C).

Heat, temperature and moving particles

You are now going to learn what is actually happening when the temperature of a substance rises.

We think that solids, liquids and gases are made up of very tiny particles, too small to be seen.

These particles are always moving. Since they are moving they have **moving energy** or **kinetic energy**. Rapidly moving particles have more moving energy than particles which are moving slowly.

Remember that particles in solids are very close together and move backwards and forwards around the same position. Particles of a liquid are free to move slightly further and are slightly less close together. Particles of gases are far apart and move long distances in all directions.

In a solid, the particles are moving slowly. In a

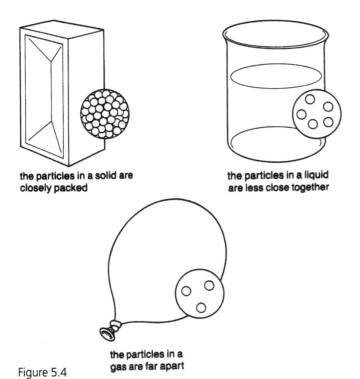

the particles in a solid are closely packed

the particles in a liquid are less close together

the particles in a gas are far apart

Figure 5.4

liquid, they are moving a bit faster, and in a gas, they are moving very fast.

When a substance or object is hot, its particles move rapidly. The particles of a cooler substance move less rapidly and the particles of a cold substance move slowly.

As a solid becomes hotter and hotter, its particles move faster and faster until eventually they are moving fast enough for the solid to become a liquid. When you heated water, its particles moved faster and faster until eventually they became fast enough for the water to boil and turn into a gas.

So, the particles of a hot substance or object have *more moving energy* than the particles of a cooler one. The

hotness or **coldness** of a substance or object is called its **temperature**.

The temperature of a substance or object shows the amount of moving energy each particle of that substance or object has.

You can see now that if you have a substance at a certain temperature, it does not matter how big a volume or mass of that substance you have. A small block of ice has the same temperature as a large block of ice. A small pot of boiling water has the same temperature as a large pot of boiling water. However, more heat energy has to be transferred to the large pot of boiling water to make all its particles move fast enough to boil.

All the objects in Figure 5.5 in a room are at 28°C. Temperature does not depend on mass or type of substance. However, the objects do *not* contain the same amount of heat energy.

Did you know that absolute zero is the coldest temperature that can exist in theory? Absolute zero is about minus 273°C. At this temperature the particles would be almost still. Nobody has ever reached absolute zero in practice, but some experiments have come very close to it.

5.2 The transfer of heat energy

What do you feel when you pick up a warm cup of tea? If you don't have a thermometer, how do you find out if someone has a fever? What do you feel if you get into unheated water on a cold day?

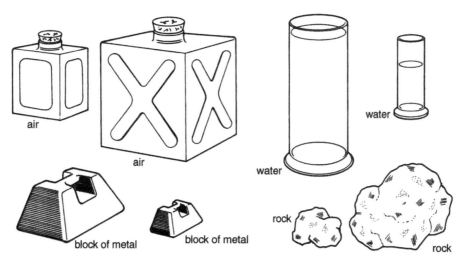

air

air

block of metal

block of metal

water

water

rock

rock

Figure 5.5

Have you ever tried cooling a hot cup of tea by placing it in a large container of cold water? Why does the tea get cooler? What happens to the temperature of the cold water?

ACTIVITY 5.2 HEAT TRANSFER

You are going to investigate the transfer of heat to and from containers of water at different temperatures. The whole class can do this as a group activity.

You will need glass beakers, glass troughs, tap water, hot water, cold water from a refrigerator, a thermometer and a watch.

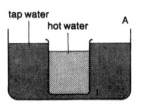
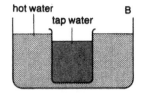
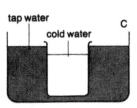
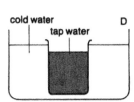

Figure 5.6

You are going to examine heat transfer in each of the arrangements (A–D) shown above.

1. In each case take the temperature of the water in the beaker and the water in the trough. Record the temperatures in your book.

2. Take the temperature of the water in the beaker and in the trough every three minutes, recording your results.

3. Leave at least one of the experiments overnight, and measure the temperature of the water in the beaker and the water in the trough the next day. Record your results.

   In each case, heat has been transferred: from where to where?

   Can you infer any general rule about the direction of heat transfer?

4. Stand a beaker of hot water in a room. Record the temperature of the air in the room and of the water

in the beaker. Record the temperature of the water after five minutes, after half an hour and after one hour. Draw a graph to show your results; put time on the horizontal axis and temperature on the vertical axis. Explain your observations.

5. Repeat step 4 using cold water from a refrigerator instead of hot water.

When a hot object or substance is in contact with another cooler object or substance, energy moves from the hot object to the cooler one. This energy that is transferred is called **heat energy**. Heat is therefore the energy transferred when there is a difference of temperature.

Of course the transfer of heat energy from the hotter substance will cause the temperature of the hotter substance to go down. This is why you feel cold if you jump into cold water. At the same time the temperature of the cooler substance will rise. The transfer of heat energy will stop when both substances are at the same temperature.

Heat energy can be transferred in three ways: by **conduction**, **convection** and **radiation**. You will learn about these in the sections that follow.

5.3 Conduction

If you walk barefoot on hot sand your feet feel hot. This is because heat energy travels to your feet by **conduction**. The hot sand is at a higher temperature than your feet. Heat energy travelled from the hotter sand to your cooler feet when they came into contact.

If you place a metal spoon in a pot containing boiling water, and then touch the spoon, what will the spoon feel like? The hot water is at a higher temperature than the spoon. So heat moved from the hot water

Figure 5.7

to the cooler spoon when they came into contact with each other. Heat then moved along the spoon from the hotter part in the water to the cooler handle.

Similarly, if you place a metal rod or spoon directly in a flame, after a little while the end of the rod or spoon becomes hot. Can you explain this? For conduction to occur there must be *direct contact* between objects and there must be a temperature difference between the objects.

 ACTIVITY 5.3 CONDUCTION ALONG A METAL SPOON

You are going to investigate how long it takes for heat to travel along a metal spoon.

You will need a metal spoon, some candle wax, a clamp and retort stand, a Bunsen burner and a watch.

1 Arrange the spoon, retort stand and clamp as shown in the diagram. Put a small lump of candle wax in the bowl of the spoon.

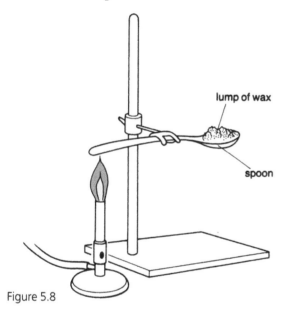

lump of wax

spoon

Figure 5.8

2 Heat the handle of the spoon gently and look at the candle wax.
How long does it take for the candle wax to start to melt?
Why does it start to melt?

The wax starts to melt because heat has been conducted from the handle to the bowl of the spoon.

 # 5.4 Conductors and insulators

Heat moves more rapidly through some materials than through others. Heat moves rapidly through metals, so metals are called **good conductors of heat**.

Heat moves slowly through some substances, and they are called poor conductors of heat, or **insulators**. Wood, plastic, glass and rubber are all poor conductors of heat.

 ACTIVITY 5.4 IS WATER A GOOD CONDUCTOR OF HEAT?

You are going to find out whether water is a good conductor or a poor conductor of heat.

You will need a large test tube, a retort stand and clamp, a piece of steel wool, water, ice and a Bunsen burner.

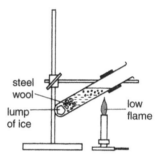

steel wool

lump of ice

low flame

Figure 5.9

1 Set up the apparatus as shown in the diagram.

2 Heat the upper part of the water over a low flame. Observe what happens to the water at the top of the tube and to the ice.
You should have seen that when the water boiled at the top of the tube the ice had still not melted. It took a long time for the ice to melt.

Answer these questions in your notebook:
a Why had the ice not melted when the water at the top of the tube was boiling?
b Why do you think the steel wool was placed above the ice?
c From your observations, what can you infer about how well water conducts heat?

This activity should have shown you that water is a

poor conductor of heat. In fact, liquids and gases in general are poor conductors of heat.

Using conductors and insulators

When we cook our food, we want heat from the flame or burner to be conducted quickly to the food. So pots, kettles and cake pans are made of metals such as aluminium, copper and iron.

Although pots and pans are made of metal, their handles are often made of hard plastic or wood. Plastic and wood are insulators and will not allow heat to flow from the hot pot to your fingers.

Figure 5.10   *Wooden and plastic handles*

Metals are good conductors of heat and good conductors of electricity. Glass and wood are poor conductors of heat and poor conductors of electricity. Solids that are poor conductors of heat are usually also poor conductors of electricity.

Some clothes are insulators

What kind of clothes do you wear when the weather is cold? What kind of clothes do you wear when the

Figure 5.11   *Woollen clothes and a cotton string vest*

weather is hot? Why do you think some clothes are warmer than others?

Woollen clothing keeps the body very warm because of the air trapped between the woollen fibres. The trapped air acts as an insulator. A blanket also prevents heat from leaving a person's body because it traps air between its fibres. A string vest traps air in its many holes and keeps the body warm.

Figure 5.12   *Vacuum flasks and a cool bag*

Materials such as fibreglass and expanded polystyrene are very good insulators because they contain a lot of trapped air. A cool bag is lined with expanded polyurethane.

On your own

1 Explain why it is better to use a wooden spoon than a metal one when stirring hot food.

2 Why does an electric iron have a metal base, but a plastic handle?

3 Get pieces of each of these substances: a ceramic tile; metal; wood; plastic. Place them on a table.
The objects are all at the temperature of the surroundings (which might be about 28°C). The temperature of your body is about 37°C. Touch each object with your hand and note how each one feels.

Which one feels coldest?
Which one feels warmest?
Can you explain why?
Which of the substances are good conductors of heat?
Which are poor conductors?

5.5 Convection

Convection is the second way in which heat may be transferred. When you boil water in a pot, you may notice that the water seems to be moving about before the water actually boils. You are going to do an activity to see how water moves when it is heated.

ACTIVITY **5.5** **CONVECTION CURRENTS IN WATER**

You are going to use potassium manganate(VII) (potassium permanganate) in order to see the currents in water that is being heated.

You will need a 250 ml beaker, a tripod, a pipe clay triangle, a glass tube, a Bunsen burner, water and a few large crystals of potassium permanganate.

❶ Pour water into the beaker until it is three-quarters full and place it on the pipe clay triangle on the tripod.

❷ Place the tube in the water near the edge of the beaker as shown in Figure 5.13 and drop two potassium permanganate crystals down the tube.

❸ Put your thumb over the end of the tube and carefully lift the tube out of the water. Try not to disturb the water.

❹ *Gently* heat the corner of the beaker where the crystals are. Write down what you see in your note book. Draw a diagram to show how the streak of colour moves in the water. The streak of colour shows how the water is moving in the beaker.

Why did you use potassium permanganate crystals rather than, say, crystals of common salt?

The movement occurs in the water because the hot water at the bottom rises and the cooler upper layer sinks and replaces it at the bottom. The heat energy in the water at the bottom moves upwards by the movement of water itself.

Heat transfer by moving material is called convection. The streams of warm moving liquids are called **convection currents**. Liquids and gases transfer heat by convection (see Figure 5.14).

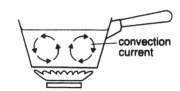

Figure 5.14

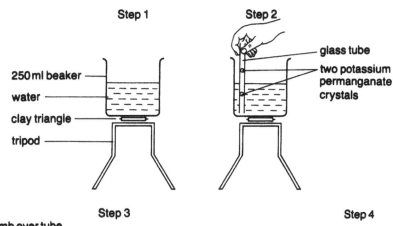

Step 1
250 ml beaker
water
clay triangle
tripod

Step 2
glass tube
two potassium permanganate crystals

Step 3
thumb over tube
some coloured water in tube

Step 4
low flame

Figure 5.13

95

ACTIVITY 5.6 CONVECTION CURRENTS IN AIR

You are going to see how convection currents in air can make things move.

You will need a circular piece of stiff card 10 cm in diameter, a retort stand and clamp, a Bunsen burner and a piece of string.

1 Draw a spiral on the card and cut along the spiral.

2 Suspend the spiral over a Bunsen burner as shown in the diagram and light the Bunsen burner with a low flame. Describe what you see.
Why does the spiral turn?

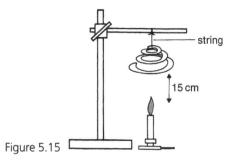

Figure 5.15

Land and sea breezes

Land and sea breezes are caused by convection currents set up in the atmosphere. If you live near the sea, you may have noticed that during the day the breeze usually blows onto the shore from the sea, but at night the breeze usually blows out to the sea.

During the day the land is heated to a higher temperature than the sea. The warm air rises above the land. Cool air from the sea replaces the warm rising air, so a breeze moves towards the land.

At night the opposite happens. The land cools faster than the sea. The sea is therefore warmer than the land. Warm air all over the sea rises and the cooler air from the land moves out to replace the warm rising air.

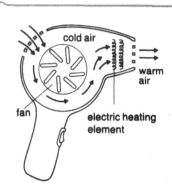

Figure 5.17

An electric hair dryer transfers heat by convection. It has a fan which pulls in air at the side and blows it over an electric heating element. What the hair dryer really does is to produce a strong forced convection current of hot air.

5.6 Radiation

If you stand in the sun or in front of a fire, you feel the heat coming from them. The heat travels from these two sources by **radiation**. The heat is not transferred by conduction because there is no contact between the source of heat and you. It is not transferred by convection because heat transferred by convection moves in an upward direction.

Heat radiation does not need any of the states of matter for its transfer. It is carried by means of waves, and can move through empty space, or a **vacuum**.

All objects give off, absorb or reflect (bounce back) radiant heat. But objects such as the sun, fires and light

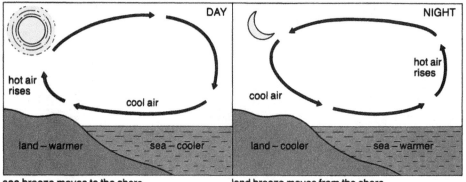

Figure 5.16

Figure 5.18   *Heat from the Sun travels by radiation*

bulbs give off a great amount of radiant heat. Objects with dull, dark surfaces absorb a great amount of radiant heat, while those with shiny surfaces reflect radiant heat. Can you explain why people who wear dark clothes in hot weather often feel uncomfortable?

Figure 5.19   *Heat travels to this girl by radiation*

The word 'radiation' comes from the Latin word for the spokes of a wheel. We still use this Latin word when we speak of the 'radius' of a circle. Just as spokes go out from a wheel's centre in several directions, so heat radiates from a point in all directions.

5.7 Temperature and our bodies

How do our bodies respond to changes in temperature?

When it is hot

Our bodies need to be kept at a steady temperature of about 36°C. When our surroundings get too hot,

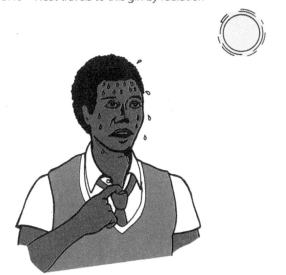

Figure 5.20   *Our bodies have to cope with both hot and cold surroundings*

changes take place to prevent the body from increasing in temperature:

a   The body produces sweat. Sweat comes to the surface of the skin and evaporates. As it evaporates, it takes heat from the body, causing the body to be cooled.

b   More blood flows near the surface of the skin. Heat from the blood is lost by radiation.

When it is cold

a   The hairs on our skin stand erect – we call these 'goose pimples'. The hairs trap air which is a poor conductor of heat. The air prevents our bodies from losing heat.

b   We also shiver because of muscular contraction and relaxation. This releases heat energy within our bodies.

5.8 Some effects of heat energy

We now know that heat energy moves from warmer bodies to cooler ones in three ways: conduction, convection and radiation. When heat energy is transferred to a body or substance, it can cause one or more of the following changes:

a   a change in temperature,
b   a change in state,
c   a change in volume (expansion).

We are going to look at the third of these changes.

ACTIVITY 5.7   THE EFFECT OF HEAT ON A SOLID

You are going to investigate the effect of heat on the volume of a solid. Your teacher can demonstrate this activity to the whole class.

You will need a metal ball and ring apparatus and a Bunsen burner.

1   Check that the ball passes through the ring.

2   Heat the ball in the Bunsen flame for about five minutes.

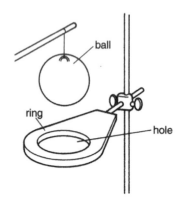

Figure 5.21

3   Now try to pass the ball through the hole.
What do you observe?
Why can the ball no longer pass through the ring?
What inference can you make about the effect of heat on a solid?

ACTIVITY 5.8   THE EFFECT OF HEAT ON A LIQUID

You are going to investigate the effect of heat on the volume of a liquid.

You will need a conical flask, a one-holed stopper fitted with a glass tube, some coloured liquid, a trough, hot water and a marker. The whole class can do this as a group activity.

1   Fill the flask completely with the coloured liquid and fit the stopper into the neck of the flask. Notice that some of the coloured liquid rises up the glass tube.

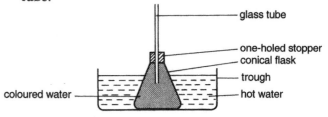

Figure 5.22

2   Mark the level to which the coloured liquid has risen.

3   Place the flask in the trough of hot water and leave it for about four minutes. Note the level of the coloured water in the glass tube.
Why has the liquid risen further up the tube?
What inference can you make about the effect of heat on a liquid?

ACTIVITY 5.9 THE EFFECT OF HEAT ON AIR

You are going to investigate the effect of heat on the volume of air.

You will need a round-bottomed flask, a one-holed rubber stopper fitted with a glass tube, a beaker, some water, a retort stand and clamp and a Bunsen burner.

1 Set up the apparatus as shown in the diagram.

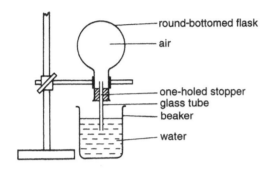

round-bottomed flask
air
one-holed stopper
glass tube
beaker
water

Figure 5.23

2 Warm the flask gently using a low flame.
What do you see coming out of the glass tube?
Why does this happen?
What inference can you make about the effect of heat on air?
Can you make the same inference about the effect of heat on all gases?

In the experiments you have done, you have used only a few different substances. You would have to do many more experiments to be certain that heat *generally* makes solids, liquids and gases **expand**. However, scientists have carried out many such experiments and they have found that this is the case.

The substances expand because their particles take in heat energy and begin to move faster and further apart. When the substances are cooled again, i.e. heat energy is taken away, the reverse happens and they **contract**. Water behaves differently from most substances: it expands on freezing.

Some effects of expansion

You have seen that materials usually expand when they are heated. The fact that most liquids expand when heated is made use of in the simple liquid-in-glass thermometers, e.g. mercury or alcohol. You were introduced to these thermometers in Book 1 (Section 3.8) and you have used the mercury thermometer in the laboratory earlier in this unit. (There is one liquid which behaves oddly as it is cooled down to close to its freezing point. Over a limited range of temperature water expands, rather than contracts, as it is cooled.)

The fact that solids expand when heated is made use of in the bimetallic strip, which is often used in thermostats. These control temperature-changing devices, such as air conditioners which keep the temperature of a room constant. Devices such as electric irons also have thermostats. A bimetallic strip consists of *equal* lengths of two different metals. The strips of metal are joined together by rivets. This prevents the strips from moving separately.

If the bimetallic strip is to work effectively the two metals chosen must expand very differently when heated through the same range of temperature. For example, if iron and copper are chosen, the copper strip expands more than the iron when heated. In order to accommodate this, the strips have to bend – see the diagram below.

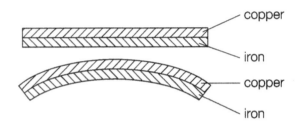

copper
iron
copper
iron

Figure 5.24

Other effects of expansion include the following:

1 Some of the older railway systems used to leave a gap between rails to allow for expansion, although more modern systems make use of single welded rails.

2 Roads are under continual stress, not only from the traffic on them, but also because of the temperature changes that may occur. The materials of the road surface and the road foundations must be such as to ensure that repeated expansion and contraction do not lead to cracking in the road.

3 Although you probably use glass items, such as beakers made out of a toughened glass like Pyrex, you often have to be careful when using glass containers. If these are made of ordinary soda glass, and you introduce a very hot liquid, the sudden expansion will cause the glass container to crack or shatter.

Heat energy

Early theory about heat

You will often read about theories in science. Scientists observe and do investigations. When they explain something that they have observed they give their explanation as a theory.

Long ago, scientists used to give explanations and develop theories by discussing and arguing with each other about what they believed took place when they observed something. Nowadays scientists still argue, but there is one major difference between the way the early scientists developed theories and the way modern-day scientists develop theories. What is this difference?

Scientists long ago thought that heat was 'a thing' called 'caloric'. They believed that caloric was invisible, it had no mass and it had no taste or smell. Now, if something is invisible it would have been difficult to prove in those days whether it existed or not.

Later on, a very interesting scientist called Benjamin Thompson, who later became known as Count Rumford, showed that the caloric theory could not account for the production of heat in some situations. He worked in Germany as a superintendent of a factory where cannons were made. As a scientist, he was observing and doing experiments. He observed that when brass was bored to make cannons, a huge amount of heat was produced. He was even able to boil water from the heat produced in boring a piece of brass. The supply of heat seemed to be never-ending as the brass was being bored. Rumford explained that the heat produced was due to the movement of the borer. His work led to the development of a theory that explains heat as a form of energy.

Kinetic theory

We know that all substances are made up of particles called molecules. These molecules have kinetic energy (energy due to movement) and potential energy (stored energy due to their position relative to other molecules). The sum of the kinetic energy and the potential energy of the molecules in an object is the **internal** energy of the object. This internal energy is also called **thermal energy**.

Thermal energy is slightly different from heat energy. The energy *transferred* between objects at different temperatures is called 'heat energy'. The energy of the molecules *within* an object is called 'thermal energy'.

Producing thermal energy

Other forms of energy can be converted into thermal energy.

Mechanical energy: In Figure 5.25, movement of one part against another produces thermal energy.

Chemical energy: When something burns, a chemical reaction takes place and thermal energy is produced. The digestion of food also involves chemical reactions which give thermal energy.

Electrical energy: We use many electric appliances in which electrical energy is converted to thermal energy.

Figure 5.26

Nuclear energy: Nuclear reactors produce a large amount of thermal energy. One of the safety precautions taken in a nuclear reactor is the use of substances to absorb some of this thermal energy.

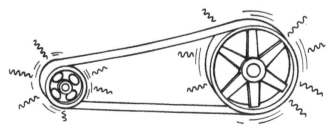

Figure 5.25

Change of state

Thermal energy is involved when a substance changes from one state to another.

ACTIVITY 5.10 MELTING ICE

You are going to measure the temperature of ice as it melts.

You will need a thermometer, a beaker, water and some ice.

1 You will be examining the temperature reading on the thermometer as the ice melts and for about five minutes after the ice melts. Prepare a table in your note book to record your results.

2 Set up the apparatus as shown in the diagram.

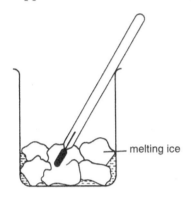

Figure 5.27

3 Record your temperature readings in your table. What time intervals are you using? One minute? Two minutes?
Explain what you observe.

4 Show your results in a graph.

ACTIVITY 5.11 BOILING WATER

You are going to measure the temperature of water as it boils.

You will need the apparatus shown in Figure 5.28.

You are going to measure the temperature of the water before, during and after it boils. Repeat the steps in Activity 5.10 but this time with water as it boils.

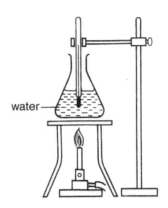

Figure 5.28

Hidden, or latent, heat

When the ice melted, it was receiving heat energy from the surrounding air because the air was at a higher temperature. But the temperature of the ice did not rise while it was melting.

As a substance is melting its temperature remains steady. This happens even if you are heating the substance to help it melt.

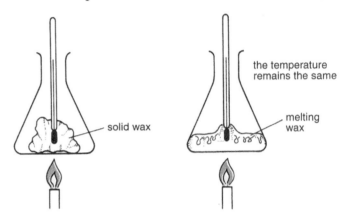

Figure 5.29

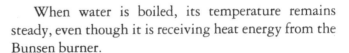

Did you know that evaporation through perspiring is a means of controlling the temperature of the human body? When someone perspires, sweat on the skin evaporates – it changes state from liquid to vapour. As it does so, it absorbs heat from the body. As a result, the body loses thermal energy and the body temperature drops.

When water is boiled, its temperature remains steady, even though it is receiving heat energy from the Bunsen burner.

When a substance is melting or boiling, the temperature reading remains the same. What is happening to the heat energy which is being supplied? The heat

101

energy is not being used to raise the temperature of the substance. Instead, the energy is being used up to **change the state** of the substance.

The heat energy used up when a substance melts is called the **latent heat of fusion**.

The heat energy used up when a substance boils is called the **latent heat of vaporization**.

When the reverse changes take place, the latent heat of fusion and the latent heat of vaporization are given up by the substance. So, when steam condenses it gives up latent heat of vaporization.

ice $\xrightarrow{\text{melts}}$ water   latent heat of fusion taken in

water $\xrightarrow{\text{freezes}}$ ice   latent heat of fusion given out

water $\xrightarrow{\text{evaporates}}$ steam   latent heat of vaporization taken in

steam $\xrightarrow{\text{condenses}}$ water   latent heat of vaporization given out

The latent heat of vaporization of water is 2260 kilojoules per kilogram. Water boils at 100°C. Suppose you have 1 kg of water at room temperature, say 25°C. To boil that water you will need to supply heat energy to raise the temperature of the water from 25°C to 100°C. Then, at 100°C you will need to supply a further 2260 kJ of energy to convert the water to steam.

Did you know that pigs wallow in mud in order to keep cool? They do this because they have no sweat glands to allow them to perspire. Instead they become cool when water in the mud evaporates from their skin.

Figure 5.30

Raising and lowering temperatures

You have just seen that when heat energy is transferred to a substance at its melting or boiling point, it becomes thermal energy inside the substance and is used to change the state of the substance.

When heat energy is transferred to a substance *not* at its melting or boiling point, the energy also becomes thermal energy inside the substance. But this thermal energy is used to raise the temperature of the substance.

We can look at this the other way around: when the temperature of a substance is raised or lowered, heat energy is absorbed or released. The amount of heat energy transferred depends on three things:

1 The mass of the substance:
The larger the mass of the substance, the more energy is involved. For example, to raise the temperature of 10 kg of water by 1°C will require 10 times the amount of energy required to raise the temperature of 1 kg of water by 1°C.

2 The temperature change:
The greater the temperature change, the greater the energy involved. For example, to raise the temperature of 1 kg of water through 10°C will require 10 times the amount of energy needed to raise the temperature of 1 kg of water through 1°C.

3 The type of substance:
The energy involved in raising or lowering the temperature of a substance differs from one substance to another. For example, about 650 J is required to raise the temperature of 1 kg of glass by 1°C and about 100 J is required to raise the temperature of 1 kg of pure iron by 1°C. The property of substances which causes this difference is called the **specific heat capacity** of the substance.

The specific heat capacity of a substance is the quantity of heat energy required to raise the temperature of 1 kg of substance by 1°C.

Water has a specific heat capacity of 4200 joules per kilogram per degree Celsius ($J \, kg^{-1} \, °C^{-1}$). This means that to increase the temperature of 1 kg of water by 1°C, it must receive 4200 J of heat energy. If its temperature drops by 1°C, it will give out 4200 J.

When we know the specific heat capacity of a substance we can calculate the heat energy involved in raising or lowering the temperature of a given mass of a substance.

Example
Calculate the heat energy released when the temperature of 4 kg of water drops from 5°C to 3°C. The specific heat capacity of water is 4200 J kg^{-1} °C^{-1}.

$$\text{temperature change } (\Delta T) = 5°C - 3°C$$
$$= 2°C$$

(Δ is the Greek letter delta, often used to indicate change. T represents temperature.)

When the temperature of 1 kg of water is lowered by 1°C,
heat released = 4200 J
Therefore, when the temperature of 4 kg of water is lowered by 2°C,

$$\text{heat released} = (4200 \times 4 \times 2) J$$
$$= 33\,600 \text{ J}$$

In general, to find the heat energy released or absorbed, we use the equation:

$$\text{heat energy} = \text{specific heat} \times \text{mass} \times \text{change in}$$
$$\text{capacity} \qquad \text{temperature}$$

On your own

1 Calculate the heat energy released when 2 kg of silver cools from 80°C to 25°C. (The specific heat capacity of silver is 240 J kg^{-1}°C^{-1}.)

2 Calculate the heat energy required to raise the temperature of 4 kg of lead by 25°C. (The specific heat capacity of lead is 130 J kg^{-1}°C^{-1}.)

Summary

Here are some of the things you have learnt in this unit:

- Heat energy is energy transferred from a body at a higher temperature to a body at a lower temperature.

- The temperature of a body (substance) is determined by the moving (kinetic) energy of its particles.

- The temperature of a body is measured using a thermometer.

- Heat may be transferred in three ways – conduction, convection or radiation.

- Heat moves more rapidly in some materials than in others. Those materials through which heat flows rapidly are called good conductors of heat.

- Metals are good conductors of heat.

- Those substances through which heat travels less rapidly are called poor conductors of heat or insulators.

- Wood, plastic, paper, cloth, liquids and gases are examples of poor conductors/insulators.

- The gain of heat energy causes a substance to a) expand, b) increase in temperature.

- A loss of heat energy causes the opposite to happen.

- There are advantages and disadvantages when materials expand.

- When a substance changes state (e.g. melts or boils) heat energy (latent heat) is used to cause the change of state.

- If ice melts to form water, latent heat of fusion is taken in.

- Different amounts of energy are needed to heat up the same mass of different substances; this difference is measured by the specific heat capacity, defined as the quantity of heat energy required to raise the temperature of one kg of the substance by 1°C.

Q U E S T I O N S

1 When mercury in a glass thermometer is immersed in melting ice, the length of the mercury in the stem of the thermometer is 2 cm. When the thermometer is immersed in steam, the length of the mercury in the stem is 17 cm. The thermometer is then immersed in liquid M and the length of the stem is 11 cm. The temperature of liquid M is

A 27°C
B 56°C
C 60°C
D 65°C.

2 The density of a solid *decreases* when it is heated because

A its mass decreases
B its volume increases
C it changes state
D the kinetic energy of the molecules increases.

3 A metal spoon and a wooden spoon are both at room temperature. When you touch each spoon, the metal spoon feels a little cooler than the wooden spoon. This is because

A the density of the metal spoon is greater
B the wooden spoon reflects heat more readily
C the metal spoon is a better conductor of heat
D the metal spoon can be polished.

4 Mercury boils at 327°C. Heat has to be supplied continuously to boiling mercury to cause it to form mercury vapour because

A when a substance changes state, there is latent heat involved.
B the kinetic energy of the liquid molecules is too low
C the vapour is a better conductor of heat than the liquid
D the boiling point of mercury is higher than that of water.

5 The temperature of a 10 kg mass of copper (specific heat capacity 400 J kg^{-1} °C^{-1} is raised from 25°C to 35°C). The amount of heat received by the 10 kg mass of copper is

A 15 000 J
B 20 000 J
C 40 000 J
D 50 000 J.

6 You are provided with three Pyrex beakers of the same size and shape. One beaker is painted black, the second is painted with a glossy white paint, and the third with a dull white paint. Each beaker is then filled with the same volume of boiling water. If you are asked to measure the temperature of the water in each beaker over an extended period of time (20–30 minutes), what would you expect to observe? Give reasons for your answer.

7 Study the diagram of the vacuum flask shown below.

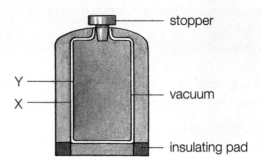

Figure 5.31

(i) If hot liquid is poured into the flask and the stopper inserted, the inside wall Y will get heated. Why are the walls X and Y silvered, and how does this affect the storage of the liquid?

(ii) If an ice-cold liquid is poured into the flask and the stopper inserted, the liquid will eventually warm up to room temperature. What are the possible sources of heat which could reach the liquid in the flask?

Light and sound energy

OBJECTIVES

- Understand the meaning of the terms light ray and light beam
- Perform experiments to show that light travels in straight lines, and how shadows are formed
- Perform experiments to illustrate reflection and refraction
- Perform experiments to show the composition of white light, and the mixing of colours
- Understand the differences between convex and concave lenses, and perform experiments to show how images are formed by convex lenses
- Describe the structure of the human eye and how defects of the eye may be corrected by use of appropriate lenses
- Describe the movement of light in terms of waves (transverse) and contrast this with sound waves (longitudinal).
- Understand ways in which the quality of sound may be changed
- Compare the speed of sound with that of light

Have you ever noticed that objects which are partly under water look strangely positioned under the surface?
In this unit, you will study how light waves travel, and investigate the phenomena of reflection and refraction. Lenses can be used to form images of objects, and for correcting faults in the human eye. Sound waves are caused by objects vibrating.

Figure 6.1 *These steps look strange under the surface of the water*

6.1 How does light travel?

We see an object when light travels from the object and enters our eyes. Some objects, such as a candle or the sun, make their own light and are called **luminous sources**. Most things we see do not make their own light but **reflect** it from a luminous source. These are called **non-luminous objects**.

It seems reasonable to suggest that light travels in straight lines – we certainly cannot see around corners! If you have a luminous source such as a candle, you can draw a diagram of the way in which light travels out from this luminous source.

Figure 6.2

You can see a non-luminous object, such as a chain, if you have a luminous source such as a lamp. The chain reflects light from the lamp, and some of this light enters the eyes of the person holding the lamp.

If we accept that light travels in straight lines, we can say that these lines are **rays of light** or **light rays**. More precisely, we can define a light ray as the direction of the path in which light is travelling. When we draw a diagram, we show a light ray as a straight line with an arrow on it.

We do not see individual light rays, but we can see **beams of light**. Think, for example, of the light from a flashlight or from a film projector. A light beam is a stream of light and is shown in diagrams as a number of rays.

A light beam may be parallel, spreading out (diverging) or getting narrower (converging).

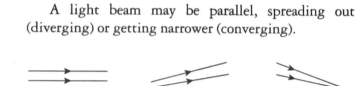

Figure 6.4

6.2 Shadows

Making shadows

Why are windows made of glass and not wood? When some objects are in the path of light rays they allow light to pass through. They are said to be **transparent**. Other objects block the light or prevent it from passing through. These objects are said to be **opaque**. When an opaque object is in the path of light, a shadow is formed. Look at the picture in Figure 6.5.

The tree is an opaque object in the path of light from the sun, so a shadow is formed. The sun changes position during the day. As the direction from which the sun's light reaches the tree changes, the shadows change in direction, length and shape. There are shadows formed by the tree early in the morning, at

light travels from the lamp to the eyes

Figure 6.3

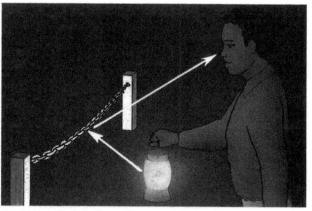

the chain can be seen because light travels from the lamp to the chain and then to the eyes

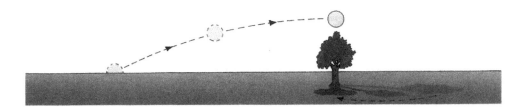

Figure 6.5

mid-morning and at midday. Can you identify which shadow was formed at the different times?

In places like the Caribbean, where it is sunny most of the time, people can look at shadows to get an idea of the time of day.

Hands make interesting shadows. Try making shadows with your hands. What happens to the shadow on a wall as you move your hands closer to the wall or further away from the wall?

ACTIVITY 6.1 FORMING SHADOWS

You are going to form shadows of an object using small and large light sources.

You will need a torch, a piece of card with a small hole in the centre, an object and a screen.

❶ To get a small source of light you can place the card with a small hole in the centre in front of the torch which has been switched on. Place the screen about 50 cm away from the torch. Place the object mid-way between the screen and the torch. Observe the shadow of the object on the screen.

Is it sharp?
Is it fuzzy?
Is it equally dark all over?

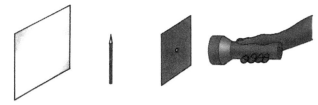

Figure 6.6

❷ Move the object closer to the light source. Describe the shadow.

How is it different from the shadow formed when the object is midway between the light source and the screen?

What does the shadow look like when you remove the card from in front of the torch?

Explaining shadows

We can show in a ray diagram how shadows are formed. The diagram below shows a shadow formed with a small light source. Remember that with a small light source you get a shadow that is equally dark all over and has sharp edges.

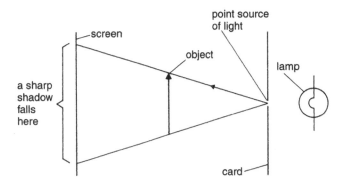

Figure 6.7

In drawing the ray diagram we can think of the source as a point since it is very small. We draw two lines representing rays coming from the point. One line is drawn touching the top of the object. The other touches the bottom of the object. The area on the screen between the rays shows where the shadow falls, because light is not getting from the light source to the screen. Light falls on the other areas of the screen, so these parts are not in shadow.

Now we will look at the ray diagram for a shadow formed with a large light source. With a large light source you get a shadow that is dark at the centre and not so dark at the edges.

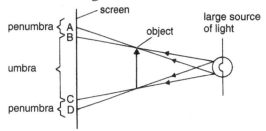

Figure 6.8

In drawing this ray diagram we cannot think of the light source as a point. Instead we take two points at

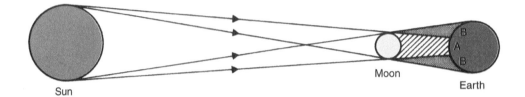

each end of the light source. From each point we draw two lines representing rays coming from these points. One line shows rays touching the top of the object, and the other shows rays touching the bottom of the object. Look at the area where the shadow falls in the diagram. Explain why the parts A to B and C to D are a partial shadow. The part of the shadow that is total shadow is called the **umbra**. The part that is partial shadow is called the **penumbra**.

Eclipses

Eclipses occur when the Earth, Sun and Moon are in a special position in relation to each other.

During a **solar eclipse**, the Moon's orbit takes it directly between the Earth and the Sun. The Sun is a large light source and with the Moon in the path of the light a shadow is formed on the Earth. The parts of the Earth in total shadow (A) experience a total eclipse. Those parts in the penumbra (B) experience a partial eclipse. In a partial eclipse part of the Sun is visible. However, do not look at a solar eclipse directly, even with dark glasses, because the light that does reach the Earth can damage your eyes.

Did you know that a total eclipse of the Sun may last for up to seven minutes? During this time the temperature falls, the sky becomes dark even though it is day, and the birds stop singing.

Sometimes the Earth can come between the Sun and the Moon. The Moon passes into the Earth's shadow and becomes almost completely dark. This is an eclipse of the Moon, or a **lunar eclipse**.

Look at Figure 6.11 and explain what happens.

Figure 6.10 *A total solar eclipse*

6.3 Reflection

When light travels to a surface like a mirror and bounces back again, we say that the light is **reflected**. When you stand in front of an ordinary flat, or **plane**, mirror you see yourself in the mirror. This is because light which travels from you to the mirror is bounced back from the mirror to your eyes. The reflection you see is called an **image**. Let us find out more about the image in a plane mirror.

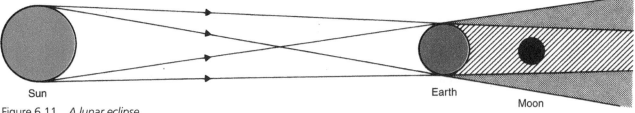

Figure 6.11 *A lunar eclipse*

1 Stand in front of a mirror. Does the image look just like you? What do you observe about the size of the image? Describe the image. Walk towards the mirror. Walk away from the mirror. What do you observe? Lift your right hand. Lift your left hand. What do you observe?

Figure 6.12 *Which hand has this boy raised? What is written on his T-shirt?*

2 You will need a small mirror and a pencil. Hold the pencil in front of the mirror. Describe the image formed. Move the pencil towards the mirror and away from the mirror. What do you observe?

3 Hold a page with something written on it in front of the mirror. What do you observe?

Describing an image in a plane mirror

Here are some facts about the image formed in a plane mirror:

1 The image is the same size as the object.

2 The image is the same distance behind the mirror as the object is in front of it.

3 The image formed is **laterally inverted**. This means that the left side of the object is the right side of the image. You can tell this when you stand

in front of the mirror and raise your left hand and then your right hand.

4 The image is **virtual**. The image only appears to be behind the mirror. We use the word virtual to describe an image that does not actually exist, but appears to exist at some place.

Light travels to form an image

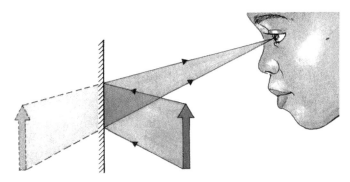

Figure 6.14

Look at the diagram above. An object is in front of the mirror. (We draw little lines on the back of the mirror to show which is the back and which is the front.) The object forms an image. This image is formed because light travels from the object to the mirror. When the light reaches the mirror it bounces back again to your eyes. The light rays come to your eyes as if they were coming from somewhere behind the mirror. It therefore appears as though there is an object behind the mirror. But in fact there is not. You only see a virtual image as *if* there was an object behind the mirror.

Reflection in other surfaces

Sometimes when you look in a pool of water you can see yourself. This happens because light from you is being reflected from the surface of the water. What you see in the water is an image, or a reflection, of yourself.

Is an image in a pool of water as clear as an image in a mirror?

Figure 6.13 *The image in a plane mirror is virtual – it doesn't really exist!*

Figure 6.15

Figure 6.16 *These are some reflecting surfaces and some non-reflecting surfaces. How are they different?*

In a pool of water not all the light is reflected. Some of the light passes through the water. When almost all the light is reflected, a clear image is formed.

On your own

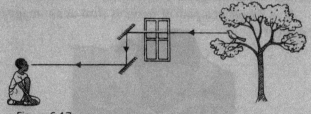

❶ Put an object between two parallel mirrors facing each other. Look in each mirror. What do you observe?

❷ Kneel below a window. Arrange two mirrors so that you can see something that is outside the window by looking at one of the mirrors at eye level.

Figure 6.17

❸ Stand two flat mirrors on their sides at right angles to each other and facing each other. Put an object, say a matchstick, between the mirrors. Look in each mirror. How many images do you see?

Using reflection

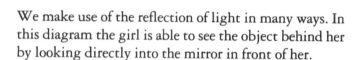
We make use of the reflection of light in many ways. In this diagram the girl is able to see the object behind her by looking directly into the mirror in front of her.

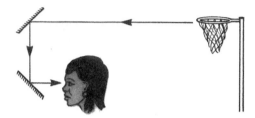

Figure 6.18

A periscope is an instrument which has a similar arrangement of mirrors. It is used in submarines to see above the surface of the water.

Car and bus drivers depend on the reflection of light to see the traffic behind them in the mirrors on their vehicles.

Driving around blind corners can be dangerous. Mirrors at the side of the road help us see the traffic around the corner.

Figure 6.19 *This mirror enables drivers to see around blind corners*

Take two mirrors. Can you use them to see the back of your head? Explain how the reflection of light enables you to do this.

6.4 Refraction

Is seeing believing?

Figure 6.20 *Water causes light to bend*

The straw in a glass of water in Figure 6.20 looks bent. Have you ever noticed this effect?

A fish swimming in a pond appears to be nearer to the surface than it really is. We observe these things because in these situations the light is behaving in a special way.

ACTIVITY **6.2** **LIGHT BENDING IN WATER AND GLASS**

You are going to investigate how light travels through water.

You will need large beakers, a coin, a ruler and some water.

1 Place the coin in the bottom of the beaker. Look at the coin and move backwards. Continue moving backwards until the coin is just out of sight. Ask a friend to pour water slowly into the beaker. Keep looking in the same direction as before.

What do you observe as the beaker fills up with water?

Figure 6.21

2 Pour water into a beaker until it is about three-quarters full. Place a ruler in the beaker, and look at it from different angles.

What do you observe?

When we look into water we can see things in the water because light from these things travels through the water and then through the air to our eyes. The light travels from one **medium** – the water – to another medium – the air. When light travels from one medium to another, the light bends. We see the light as if it is travelling in a straight line. Therefore the light from the water appears to be coming from a point that is on a straight line with the light coming to our eyes.

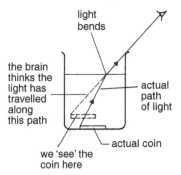

Figure 6.22

This bending of light as it travels from one medium to another is called **refraction**. It is because of refraction that a pool appears shallower than it really is, and a straw appears bent when placed in water.

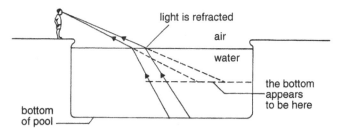

Figure 6.23

ACTIVITY 6.3 LIGHT TRAVELLING THROUGH GLASS

You are going to investigate how light travels through a glass block.

You will need a glass block, some pins, a piece of white paper and a ruler.

1. Place the glass block on the piece of white paper. Use a pencil to make a drawing of the outline of the glass block (PQRS).

2. Draw a normal (ON) and a line (OT) at an angle of about 45° with ON.

3. Place two pins (1 and 2) on the line OT, as far apart as possible.

4. Look carefully through the block and place two pins (3 and 4) in line with the images of pins 1 and 2.

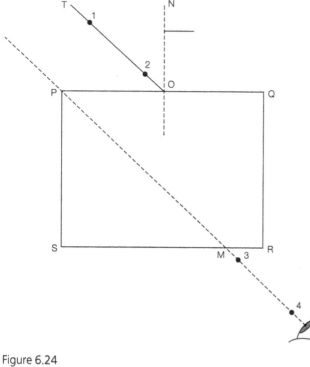

Figure 6.24

5. Take out each of the four pins and mark each spot with a small cross.

6. Remove the block. Draw a line through the crosses where there were pins 1 and 2. Make sure that the line meets the side PQ at O. The line you have drawn represents a light ray falling on the block.

7. Now draw a line through the crosses where there were pins 3 and 4. Make sure that the line meets the side RS at M. This line represents a light ray coming out of the block.

8. Join points O and M. The line OM represents the **refracted ray** in the glass block.

   Is the light ray which enters the glass block bent towards or away from the normal?
   What happens when the light ray comes out of the glass block at M?

ACTIVITY 6.4 THE COLOURS OF THE SPECTRUM

You are going to separate the light from a bulb into its separate colours.

You will need a triangular glass prism, a piece of white paper, a piece of card and an electric light bulb.

1. Put the sheet of white paper on your desk and shine the light bulb along it.

2. Cut a slit in the piece of card and place it in front of the light bulb so that a narrow ray of light travels along the paper.

3. Place the prism in the path of the ray of light as shown in Figure 6.25.

   What do you observe?

   What is the colour of the light that goes into the prism?

Figure 6.25

What is the colour of the light that comes out of the prism?

The splitting up of white light into a spectrum is called **dispersion**.

Did you know that rainbows are formed by the refraction of light from the sun in raindrops?

You have split the light from the electric light bulb (which is white or nearly white) into separate colours by refraction through the prism.

A rainbow is formed when light from the sun is refracted as it passes through drops of moisture in the atmosphere. The light from the sun is called **white light**. This light is made up of light of different colours. When white light is refracted through a rain drop each of the coloured lights bends to a different extent. This causes the colours to separate and we get a rainbow made up of the different colours. The colours are red, orange, yellow, green, blue, indigo and violet. These are called the colours of the **spectrum**. They are almost the same as the colours that emerged from your prism.

6.5 Lenses

Some instruments which make use of the refraction of light are cameras, spectacles, microscopes and telescopes. These instruments contain a part, usually made of glass, called a **lens**.

Figure 6.26 *All these objects have lenses*

When light passes through a lens it bends or is refracted. A **convex lens** causes parallel light rays to bend inwards, or **converge**. A **concave lens** causes parallel light rays to bend outwards, or **diverge**. Because of this a convex lens is sometimes called a **converging lens**. A concave lens is also called a **diverging lens**.

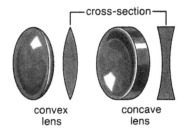

Figure 6.27 *How are these two types of lenses different?*

ACTIVITY 6.5 THE FOCAL LENGTH OF A CONVEX LENS

You are going to form a clear image with a convex lens, and measure the distance from the lens to the image.

You will need a convex lens and a ruler.

1. Hold the convex lens close to a wall in the path of light from a window. The window should be quite a long way from the lens.

2. Move the lens back and forth until you see a sharp image on the wall.

3. Ask a friend to measure the distance between the lens and the wall when you get a clear image.

4. Move the lens a little towards the wall. Move it a little away from the wall.
   What do you observe?

Did you know that a camera is like the human eye in some ways? You will find out how they are alike on page 118.

The light coming through the lens in Activity 6.5 formed an image on the wall. How was this image different from the image formed by a plane mirror?

The light passes through the lens and gives an image on the wall. When we get a clear image, we say that the image is **in focus**. The image is at the **focal point** of the lens. The distance between this point and the lens is called the **focal length** of the lens. A camera uses a lens to bring light to focus on the film at the back of the camera.

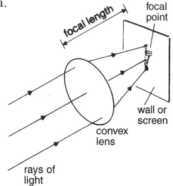

Figure 6.28 *The focal length is the distance between the lens and the focal point*

ACTIVITY **6.6** **FORMING IMAGES WITH A CONVEX LENS**

You are going to form images of objects which are at different distances from a convex lens.

You will need a source of light (e.g. a bulb with a screen with a tiny hole for a small beam to pass through), a convex lens, a lens holder, a long pin, a screen and a metre rule.

1 First find the focal length of the lens.

2 Set up your apparatus as shown in Figure 6.29.

3 Place the pin at different distances from the lens as listed below. Each time, move the screen until you get a clear image, and measure the distance from the lens to the screen.
Distances of the pin from the lens:
a twice the focal length,
b more than one and less than two focal lengths,
c more than two focal lengths.

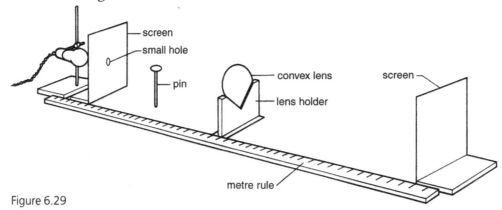

Figure 6.29

4 In each case describe the image formed.
How far is the image from the lens?
Does it look bigger or smaller than the object?
Is it the right way up?

5 Put the pin close to the lens at a distance of less than one focal length.
Are you able to form an image?

6 Record your observations in a table.

We can use a convex lens to form images. The image can be larger than the object (**magnified**), the same size as the object, or smaller than the object (**diminished**). The image may also be the right way up or upside down (**inverted**).

Looking at images formed by a convex lens

1 Object at twice the focal length.

object   image

image is the same size as object
image is inverted

Figure 6.30a

2 Object at position between focal length and twice focal length.

object   image
image is magnified
image is inverted

Figure 6.30b

3 Object at position beyond twice focal length.

object   image

image is smaller or diminished
image is inverted

Figure 6.30c

4 Object within the focal length.

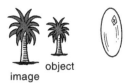

image object

Figure 6.30d

Look at the diagram of the image formed when the object is within the focal length. How is this situation different from the others?

The images formed in Figures 6.30a, b and c are said to be **real** since the rays, on passing through the lens, bend and meet to form the image. We say that the rays converge. The convex lens is also called a converging lens.

In Figure 6.30d the rays, on passing through the lens, do not meet but appear to be coming from the position where the image is formed. This type of image is similar to that formed in a plane mirror. We say that the image is **imaginary** or **virtual**.

Concave lens

When light passes through a concave lens, the rays end up away from each other or diverge. The concave lens is called a diverging lens. What type of images would you expect to be formed using a concave lens?

A magnifying glass

Look at an object through a magnifying glass. Describe what you see. Move the magnifying glass towards and away from the object. From what you have learnt about lenses, explain how the magnifying glass works.

Light and colour

In Activity 6.4 we saw that white light can be split into different colours using a prism. When light of all the colours of the spectrum is mixed together we see white light. But what makes objects look white or coloured?

Objects reflect light. When an object looks white it is reflecting all the coloured light in the spectrum. When it looks black it is absorbing all the coloured light in the spectrum. An object looks red because it absorbs all the other colours and reflects red.

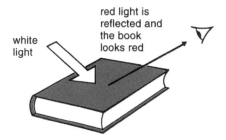

Figure 6.31   *The red book absorbs all the colours in white light except* red

Remember that light is a form of energy. Can you explain why you feel cooler when you wear a white shirt on a hot day rather than a black shirt?

If you have coloured filters you can do interesting things with coloured light.

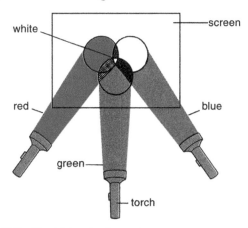

Figure 6.32   *Try shining torches onto a screen*

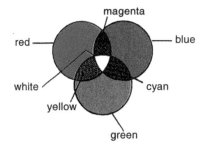

Figure 6.33   *Combining coloured lights forms other colours*

You can combine coloured light to form other colours. If you combine red, blue and green light you will get white light. These are called primary colours.

Red and blue give a colour called magenta.

Red and green give yellow.

Blue and green give a colour called cyan.

115

Scientists are not yet able to explain fully how the human eye is able to see colours. At the back of the eye we have **sensory cells** that are of two types: **rods** and **cones**. The rods become sensitive with dim light. The cones are only sensitive to bright light and they enable us to see colour.

The rods are arranged at the edge of the retina, so we use them to see objects which are not straight ahead of our eyes. The cones are at the centre of the retina, so we need to look fairly straight at objects to use them.

On your own

You are going to see where an object has to be placed for you to see its colour.

Cut some small squares of different coloured shiny papers and mix them up. Pick one up with your eyes closed and hold it out to your side with your arm out-stretched. Open your eyes and look straight ahead; you must not look to the side.

Swing your arm round slowly. You should be able to see it from the corner of your eye. At first you won't be able to see what colour it is, but at a certain position you will be able to see its colour. You will find that this position is different for different colours.

Which colour do you see first: red or blue?

Write down the order in which the colours become visible.

Do your classmates get the same results?

It is easier to see very faint stars by looking through the corner of your eye than by looking straight at the stars. This is because rods, which are at the edge of the retina, are more sensitive to dim light than cones, which are at the centre of the retina.

 6.6 The human eye

Our eyes are very precious. People who can see often take their eyes for granted. However, we need to take care of them and protect them.

The various parts of our eyes work together to enable us to see. Figure 6.34 shows the parts of the eye. The table below describes the special function of each part.

Part of the eye	Function
cornea	Transparent covering that refracts light and helps the lens focus light.
iris	A ring of muscles that controls the amount of light entering the eye.
pupil	Opening in the centre of the iris that allows light to pass through.
lens	Refracts the light rays to focus the image on the retina.
ciliary muscles	Alter the shape of the eye lens.
sclera	Tough outer layer that protects the eye (the cornea is part of the sclera).
retina	Serves as a light-sensitive screen for the formation of the image.
optic nerve	Takes messages from the cells of the retina to the brain.

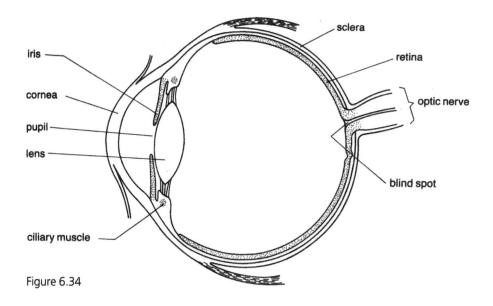

Figure 6.34

Light rays entering the eye pass through the eye lens. The light is refracted and comes to focus on the retina. The cells in the retina are **light-sensitive**. This means that when light falls on them they react and send a message to the brain by way of the optic nerve. The brain interprets it and tells us what we are seeing.

Figure 6.35

Professor Ramsey McDonald Saunders was Head of the Department of Physics at the University of the West Indies. He has researched in many different areas including the way in which the brain processes the optical signals which come along the optic nerve from the eye. He has also developed a machine to measure abnormal curving of the backbone and has been involved in many development projects.

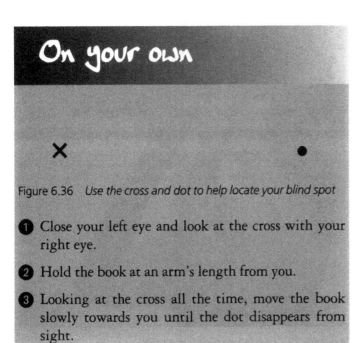

Figure 6.36 *Use the cross and dot to help locate your blind spot*

1 Close your left eye and look at the cross with your right eye.

2 Hold the book at an arm's length from you.

3 Looking at the cross all the time, move the book slowly towards you until the dot disappears from sight.

The dot has disappeared because the image of the dot has fallen onto the **blind spot** of the retina of the right eye. This is the point where the optic nerve enters the eye – there are no light-sensitive cells at that point.

With both eyes we are able to see both the cross and the dot all the time as we move the book towards our eyes. This is because when the dot falls on the blind spot of one eye, it is not on the blind spot of the other.

Adjustment of the eyes

When we look at things, our eye lens adjusts in order for us to see things that are far away and things that are close up. Our eyes also adjust for us to see things in bright light and in dim light.

ACTIVITY 6.7 ADJUSTMENTS MADE BY EYES

You are going to investigate some of the adjustments your eyes make to cope with objects at different distances and with different brightnesses of light.

1 Close your eyes. Put a hand over one eye. With the other eye, look at something that is far away from you. Turn your head quickly and look at something close to you.
What do you notice?
You may have found that it took a few moments for the thing you were looking at to become clear. You were focusing on something far away, then had to focus on something close to you. It takes a few moments for the ciliary muscles to change the shape of the eye lens so that you can focus on the close object.

2 Hold your book away from you. Move it slowly towards you until you no longer see the words clearly. Measure the distance between the page and your eyes.

3 Look straight into a friend's eyes. Ask him or her to close them for one minute. Looking straight into the eyes, ask your friend to open them when the time is up.
What do you notice?
You may have noticed that the pupils in the centre of each eye became smaller. The muscles of the iris make the pupils smaller in bright light so that less light enters the eye.

117

The camera and the eye

The camera resembles the human eye in some ways. In the eye, light passes through the convex eye lens and comes to focus on the retina. In a camera, light passes through a convex lens and forms an image on film at the back of the camera.

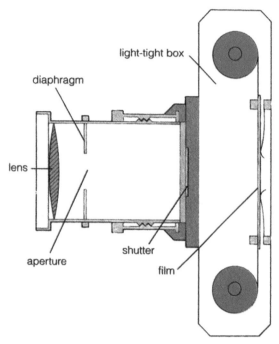

Figure 6.37

The following table lists some of the ways in which the camera resembles the eye.

Eye	Camera
lens refracts light rays	lens refracts light rays
image formed on retina	image formed on film
light enters through pupil	light enters through aperture
iris adjusts amount of light entering eye	diaphragm adjusts amount of light entering camera

In normal vision in the eye, the ciliary muscles change the shape of the lens. This allows us to see objects that are close up or objects that are far away.

In the camera a different method is used to form images of objects at different distances. The lens can be moved nearer to or further from the film, so that the image will be focused on the film.

Helping our eyes

Instruments to help us see

When we want to see small things (such as an insect) properly, we can use a magnifying glass – this is a type of convex lens. In order to see more details on very small things (such as protozoa) we use a microscope.

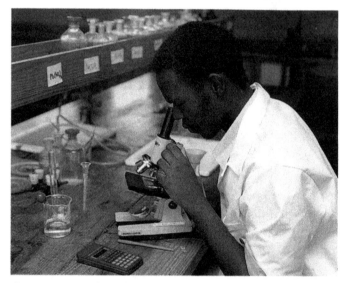

Figure 6.38 *Looking through a microscope*

When we want to see things that are far away we use binoculars: this makes the things look bigger. We can also use an instrument called a telescope. Scientists use telescopes to study stars and other things in space.

Figure 6.39 *Looking through binoculars*

Short sight (inability to see long distances) can be corrected with concave lenses.

Long sight (inability to see short distances) can be corrected with convex lenses.

Figure 6.40 *Short and long sight can both be corrected with lenses*

Problems with seeing

Some people have difficulty seeing things that are far away that other people can see clearly. People who cannot see things that are far from them, but can see things that are close to them, are said to be **short sighted**. These people can use spectacles with **concave lenses** in order to help them to see distant objects.

People who can see things that are far away, but have difficulty seeing things that are close up, are said to be **long sighted**. These people can use spectacles with **convex lenses** in order to help them to see close objects.

When the cornea in the eye is uneven, there may be problems with forming a clear image on the retina. A person with this problem is said to have an **astigmatism**. This can be corrected by using specially shaped spectacles.

Think how lucky you are if you can see this page! Blind people use several devices to help them to function independently. They make use of other senses like sound and touch to get information. Blind people can also read although they cannot see. They do so by a method known as Braille, in which instead of printed letters there are patterns of raised dots on a page. By touching the pages in a Braille book, they can feel the words that are written.

Care of the eyes

Here are some of the things that we can do to take care of our eyes:

1. Eat foods that contain vitamin A, such as yellow foods like carrots and pumpkin.
2. Read in good lighting.
3. Do not look into bright lights or the Sun. This can cause damage to the retina.

Figure 6.41 *This boy is reading using Braille*

Did you know that tears help clean our eyes? Our eyelashes also protect our eyes.

4. Do not put anything in your eyes unless a doctor tells you to.
5. If something gets into your eye, you may be able to remove it by washing the eye in an eye bath. Try not to rub your eye, because whatever is in the eye could scratch the cornea.

6.9 Making sounds

Sound is a form of energy. It is something that we can hear. Do you know what causes sound?

ACTIVITY 6.8 SOUND AND VIBRATIONS

You are going to investigate what happens when a musical instrument makes a sound. The whole class can do this activity together.

You will need some stringed and some percussion instruments such as guitars, cymbals, drums and gongs, a tuning fork and a beaker of water.

1. Pluck a string of a stringed instrument. What do you hear?
   Look carefully at the string while it is making a note. What can you see?
   Hold a piece of thin paper so that it is just touching the string. What can you see?

2. Strike one of the percussion instruments and, while it is still making a noise, hold a piece of paper on the part you hit.
   What do you see?

3. Put your hand over your mouth and, with your lips lightly together, blow so that you make a noise.
   What do you feel in your lips?

4. Strike the tuning fork so that it makes a note, and hold it so that the end of the tuning fork is just touching the water in the beaker.
   What do you see?

Figure 6.42

In all the above cases you should have detected a backward and forward movement when you heard sound. This kind of movement is called **vibration**. A sound is produced when an object vibrates. For example, when a person plucks a guitar string it vibrates and we hear a sound.

Sound makers

We make sounds when we cause an object to vibrate. Look at the sound producers below. Which parts are vibrating in order to produce sound?

a

b

c

Figure 6.43

On your own

You can make sounds with some simple materials.

1. Make a guitar with some rubber bands, pins and a box.

2. Use bottles containing water to make different sounds as shown in Figure 6.44.

3. Use a cardboard tube to make a trumpet.

4. Use tins to play a tune.

Figure 6.44

6.10 How sound travels

A sound is produced when an object vibrates. The vibrating object causes the air around it to vibrate. These vibrations ripple through the air and we get a sound wave.

If you strike a tuning fork on the table and hold it up in the air it vibrates and you hear a sound. How does this sound travel to your ear? The vibrating fork causes the air around it to vibrate. This vibrating air causes more air to vibrate. A wave is produced as the air vibrates from the tuning fork to your ear. The air vibrates backwards and forwards. Some parts of the air are compressed together and other parts are pushed away from each other.

Sound also travels through solids and liquids.

Sound will not travel through a vacuum. We can show this using the apparatus in the diagram below. The electric bell is switched on and the air from the bell jar is gradually removed by the pump. As the air is removed from the bell jar, the ringing sound gets fainter and fainter until in the end it cannot be heard at all because there is no air to vibrate.

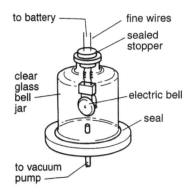

Figure 6.45

Astronauts on the Moon have to talk to each other using radio, because there is almost no air to carry sound.

On your own

You can use a spring to make a wave which is rather like the waves that sound makes in the air.

Hold the ends of a long spring with your hands. Keep one end fixed. Move the other end backwards and forwards as shown by the arrows in the diagram. Observe the spring as it vibrates. Describe the movements you see in the spring.

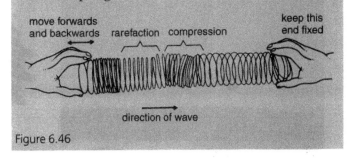

Figure 6.46

When the spring begins to vibrate, parts are compressed together (compression) and parts move apart (rarefaction). Each bit that is pressed together appears to move along the spring from one end to the other. We say that a wave is moving along the spring. This kind of wave, which travels in the same direction as the vibrating parts, is called a **longitudinal wave**.

121

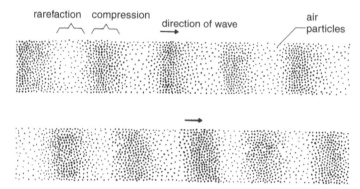

rarefaction   compression   direction of wave   air particles

Figure 6.47

Sound is caused by vibrating air particles. The air particles are pressed together and moved apart just as your spring was. A sound wave in air travels in the same direction as the vibrating air particles (see Figure 6.47). Therefore, a sound wave is a longitudinal wave.

 ## 6.11 Different kinds of sound

Loud sounds and soft sounds

Some sounds are pleasant. Others make our ears feel uncomfortable. How loud a sound is depends on how much energy the sound waves carry. This energy is called the **sound intensity**. A loud sound has a high sound intensity.

We measure sound intensity in **decibels**. The symbol for decibels is dB. A safe sound intensity for the human ear is 85 dB. A person who is exposed to high sound intensities must use ear protection devices since

Figure 6.49   *A situation of excessive noise*

high sound intensities can cause a loss of hearing. The following table gives a list of some sounds and their intensities. Excessive sound causes noise pollution. Like other types of pollution we need to take measures to reduce noise pollution in our environment.

Source of sound	Approximate sound intensity (dB)
jet taking off	150
amplified disco music	115
noisy classroom	77
busy traffic	70
conversation	60
hair dryer	50
soft whisper	30
leaves rustling	10

Sounds of low and high pitch

When a guitarist plucks the bottom string in a guitar you hear a high sound. When s/he plucks the top string you hear a low sound. These sounds differ in **pitch**. We can describe sounds as having a high pitch or a low pitch.

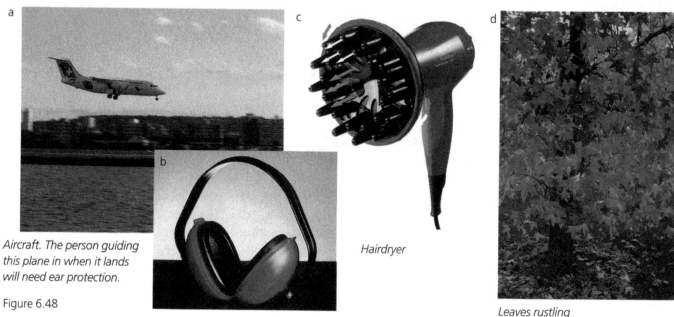

a   *Aircraft. The person guiding this plane in when it lands will need ear protection.*

Figure 6.48

c   *Hairdryer*

d   *Leaves rustling*

ACTIVITY 6.9 DETERMINANTS OF PITCH

You are going to investigate what affects the pitch of a vibrating string or wire.

You will need string and wire of different thicknesses, some weights, two triangular blocks of wood, and a nail or drawing pin to secure one end of the string or wire.

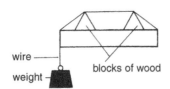

wire ——

weight ——

blocks of wood

Figure 6.50

The diagram shows you how you can set up the apparatus.

1. Find out if the length of a vibrating string or wire affects the pitch of a sound.
   What should you do?
   What should you change?
   What should you keep the same?

2. Find out if the **tension** (tightness) of the string or wire causes the sound to change.
   What should you change?
   What should you keep the same?

3. Find out if the thickness of the wire or string causes the sound to change.
   What should you change?
   What should you keep the same?

When you change the **length**, the **tension** and the **thickness** of the wire, the pitch of the sound changes.

Length	A shorter length produces a note of higher pitch.
Tension	A tenser wire produces a note of higher pitch.
Thickness	A thinner wire produces a note of higher pitch.

Change in pitch

We have seen that sounds can differ in pitch. When the guitarist plucks the bottom string it vibrates and gives a note of a high pitch. This string is vibrating very fast. It vibrates 660 times in one second. When the guitarist plucks the top string we get a note of low pitch. This string is vibrating 165 times in one second.

The number of vibrations in one second is called the **frequency**. Therefore the bottom string has a higher frequency than the top string which has a lower frequency.

The higher the frequency, the higher is the pitch of the sound that we hear.

Quality of sound

Different instruments are made of different materials and are of different shapes. Because of this, the sounds they produce are different even if the pitch is the same. For example, you would be able to tell whether a note you heard was produced by a steel pan or a guitar. We say that the notes produced by the instruments differ in **timbre**.

Wind instruments

You have seen examples of stringed and percussion instruments in your activities. There is another group of instruments, called wind instruments, into which you blow. In a wind instrument the sound is produced by a vibrating column of air. If you shorten the column of air a high pitched note is formed. How can you shorten or lengthen the column of air in a wind instrument?

Where does the air vibrate in these wind instruments?

Figure 6.51 *Wind instruments*

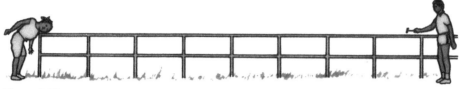

Figure 6.52

6.12 Speed of sound

Sound travels at different speeds in solids, liquids and gases. If you put your ear at one end of a metal rail and someone taps at the other end, as shown in Figure 6.52, you hear a sound. This sound travels faster through the metal rail than through the air. Sound travels most rapidly in solids, less rapidly in liquids and very slowly in most gases.

ACTIVITY 6.10 MEASURING THE SPEED OF SOUND IN AIR

You are going to measure approximately how fast sound travels in air. The whole class can do this as a group activity.

You will need two dustbin lids and a stopwatch.

❶ Go outside to a wide open space. Let one person with the two dustbin lids go as far away from the others in the group as possible where s/he can still be seen clearly. Measure the distance from the main group to the person holding the lids. You can do this using paces, if you know how long one pace is.

❷ This person should then bang together the two dustbin lids very hard. As soon as the person holding the stopwatch sees this, s/he should start the stopwatch. As soon as this person hears the sound of the dustbin lids banging, s/he should stop the stopwatch.

❸ Write down the length of time recorded on the stopwatch. Repeat this procedure a few times so that you see whether you get the same results.

Light travels so fast, that we assume we see the person banging together the lids as soon as it happens. So the time between seeing the banging of the lids and hearing the banging of the lids is the time it takes for the sound to reach us.

Let us say that the person with the lids was 700 metres away, and that the time the sound took to reach you was 2 seconds. Then the speed of sound in air would be:

speed = distance/time
     = 700/2
     = 350 metres per second.

In fact, the speed of sound in air is about 330 metres per second. The speed increases as the temperature of the air increases.

The speed of sound in steel is about 6000 metres per second.

Echoes and the speed of sound

Have you ever heard an echo of your voice? If you are in a cave or in a large church hall or auditorium, you should be able to hear an echo when you shout something. When you shout, the sound waves travel from you to the walls of the building or the sides of the cave. The waves then bounce back and travel to your ears and you hear the echo. In order to hear an echo you should

Figure 6.53

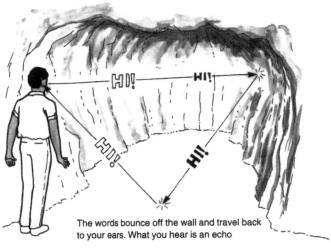

Figure 6.54

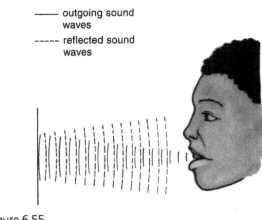

Figure 6.55

The words bounce off the wall and travel back to your ears. What you hear is an echo

be about 100 m away from the wall. From the time when you make the sound until you hear the echo, the sound travels from you to the wall and back.

Sometimes in an empty hall or in a cave you hear several echoes when you make a sound. Can you explain why this happens?

Ships use echoes to estimate the depth of water beneath them. A sound is sent downwards from the ship. The sound waves bounce back from the sea bed and the echo is received by the ship. Since the speed of sound in water is known, and the time taken for the sound to travel to the sea bed and back has been measured, the distance from the sea bed can be calculated. This way of finding depths or distances using sound is called **sonar** or **echo finding**.

Did you know that scientists use sonar in various materials to search for oil and minerals and to find objects sunk at the bottom of the sea?

6.13 Some sounds we cannot hear

You have seen that the pitch of a sound refers to the frequency of the vibrating object. Frequency is measured in units called **hertz** (Hz). This unit is named after a German physicist called Heinrich Hertz.

1 Hz = 1 vibration in one second

1 kilohertz (kHz) = 1000 Hz

The human ear can hear sounds with frequencies that range from about 25 Hz to about 20 kHz, or

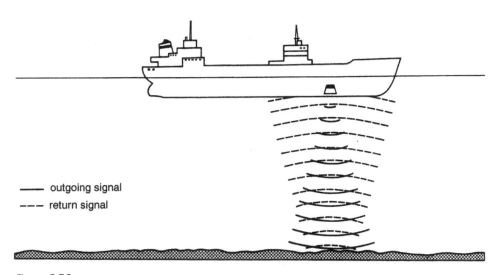

— outgoing signal
--- return signal

Figure 6.56

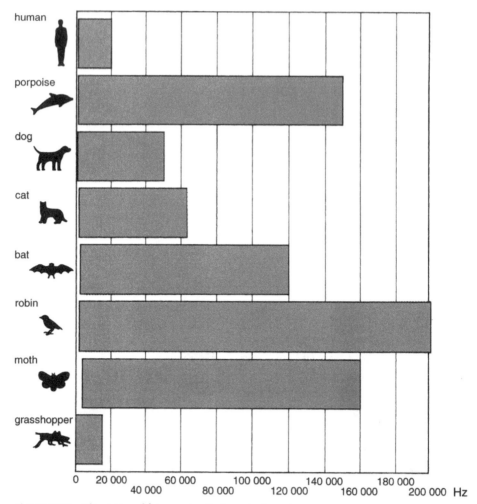

Figure 6.57   *The range of frequencies that can be heard*

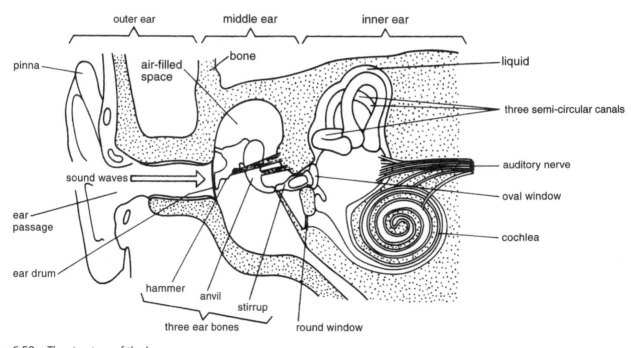

Figure 6.58   *The structure of the human ear*

20 000 Hz. Sounds with frequencies below 25 Hz are called **infrasonic**. Sounds with frequencies above 20 kHz are called **ultrasonic**. Humans cannot hear these ultrasonic frequencies but some animals can. The diagram in Figure 6.57 shows the range of frequencies that some animals can hear.

Bats make use of their ability to hear ultrasonic frequencies when they find their way about in the darkness. They emit high frequency sounds which are reflected from objects. When the bats hear the echo they can tell where the objects are.

Ultrasonic sounds are used in echo finding by ships. Ultrasonic sounds are also used in medicine to detect some types of cancer and brain damage and to study how an infant is developing in the womb.

Did you know that muscles in the human body produce sound of a frequency of about 20 Hz? By studying the sound that is made when a person lifts certain loads, scientists find out about how some muscles can be developed for certain sporting activities.

6.14 The human ear

The human ear is designed in such a way that it receives and transmits the vibrations in the air. This is what enables us to hear (see Figure 6.58).

How we hear

The **outer ear** (**pinna**) collects the sound waves (vibrations in air) and channels them down the **ear passage**. The sound waves strike the **ear drum** which is stretched across the end of the passage. The ear drum vibrates and these vibrations are transferred to the little bones in the **middle ear**: the **hammer**, the **anvil** and the **stirrup**. The ear bones vibrate and transfer the vibration across the middle ear to another membrane, the **oval window**.

Did you know that the stirrup bone in the ear is the smallest bone in the human body?

The oval window vibrates and causes the fluid in the inner ear to vibrate. A structure called the **cochlea** is suspended in this fluid. The cochlea contains sensory hairs. As the cochlea vibrates the fluid inside it vibrates and the sensory hairs vibrate. As the sensory hairs vibrate they produce nervous impulses which pass along the **auditory nerve** to the brain where they are interpreted as sound.

The sensory hairs are of varied length. The long ones are made to vibrate by small vibrations of low sounds while the shorter ones vibrate only with high sounds.

There are also structures called **semi-circular canals** in the ear. These send messages to the brain as we move and help us to maintain our balance.

Wild animals generally have much more sensitive hearing than humans. Animals such as rabbits have large ear flaps. They can turn these towards the source of a sound so that they can collect as much sound as possible.

Helping our sense of hearing

We can use instruments to help us hear better. People who are partially deaf use a hearing aid. A doctor listens to a person's heartbeat by using a stethoscope.

6.15 Acoustics

Have you ever tried to block out sound? What do you do? You can put a pillow, cushion or soft material over your ear or use ear muffs (see Figure 6.48). The pillow, cushion or soft material reduces the loudness of sounds because they are able to absorb the sound.

On the other hand, hard surfaces like a wall will cause the sound to bounce back. The surfaces reflect the sound. When sound is reflected we get an echo.

You may have noticed that sometimes when you make a sound in an empty hall you do not hear a distinct echo. Instead you hear a prolonged sound. This happens because the sound that is reflected mixes with the original sound that you made. If a sound is reflected off several surfaces a prolonged sound may be heard.

Because sound can be reflected and absorbed, a concert hall is carefully designed for the audience to hear the performers clearly. A hall or auditorium in

Figure 6.59   *A concert hall*

which sounds can be heard clearly is said to have good **acoustics**.

An empty room will not have good acoustics because the sound will be reflected and produce echoes. When curtains, rugs, furniture and cushions are in a room, some of the sound is absorbed and the acoustics improve. People who design concert halls and auditoriums need to know about the reflection and absorption of sound in order to design a building with good acoustics.

Did you know that a recording studio is sometimes called an 'anechoic chamber'? The studio is specially designed with the floors, walls and ceilings covered with sound-absorbing baffles. These baffles prevent echoing.

Summary

These are some of the main ideas you have learnt in this unit:

- Light and sound are forms of energy.

- Light travels in straight lines in all directions.

- Light can be reflected. Because of the reflection of light we see images in a plane mirror.

- Light is refracted when it passes from one medium to another.

- White light is made up of the colours of the spectrum which we see in rainbows.

- A lens is a part of many instruments that make use of the refraction of light.

- The human eye contains a convex eye lens.

- The parts of the eye help us to see.

- Long sight, short sight and astigmatism are three problems that people may have with their eyes. The use of special lenses helps sufferers to see better.

- Our eyes are important sense organs and we must take care of them.

- Sound is produced when an object vibrates; it travels as sound waves, which are different from light waves.

- Sound waves are longitudinal waves and travel by means of a series of compressions and rarefactions.

- Sound can travel through solids, liquids and gases; it cannot travel through a vacuum.

- Sound can be described by its intensity, pitch and quality.

- The unit for measuring sound intensity is the decibel (dB).

- The pitch of a sound depends on the frequency; the higher the frequency the higher the pitch.

- Frequency is the number of vibrations in one second and is measured in hertz (Hz).

- The pitch of a stringed instrument changes with the length, tension and thickness of the vibrating string.

- The same note played on different instruments has a different quality or timbre.

- The speed of sound can be calculated from measurements of distance and time.

- Sound can be reflected and absorbed. The reflection of sound causes echoing.

- The parts of the ear help us hear sounds.

- Buildings like churches and concert halls can be designed to give good acoustics.

Q U E S T I O N S

1 When a gun is fired, smoke can sometimes be seen. The smoke will be seen by an observer some hundreds of metres from the gun long **before** the sound of the firing is heard by the same observer. This difference happens because

  A   the speed of the bullet fired is greater than the speed of sound
  B   the observer's eye reacts more quickly than his ear
  C   the speed of light is greater than the speed of sound
  D   sound travels outwards from the source in all directions.

2 An echo sounder can be used by a ship to detect the depth of the water beneath the ship. When an echo sounder was used by a ship, it was found that the depth of water beneath the bottom of the ship was 400 m. If the interval of time between each sounding is 0.2 s, the speed of sound in water would be

  A   1200 m/s
  B   2400 m/s
  C   3600 m/s
  D   4000 m/s

3 A total eclipse of the sun may be seen **only if** the

  A   Earth is between the Sun and the Moon
  B   Moon is between the Sun and the Earth
  C   Sun is between the Moon and the Earth
  D   observer is on the darkened side of the Earth.

4 If a metre rule is placed in water as shown in the diagram, it will appear to an observer to be bent at the surface (O) because

  A   water is less dense than air
  B   the metre rule is bent at the surface
  C   light travels faster in water than air
  D   water is more dense than air.

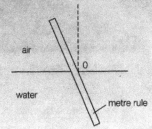

Figure 6.60

5 If the focal length of a convex lens is f, the lens will produce a same-sized, inverted image as long as the distance of the object from the lens is

  A   equal to f
  B   equal to $2f$
  C   between f and $2f$
  D   less than f

6 An object is placed 30 cm from a convex lens. The image of the object is formed 30 cm from the lens but on the opposite side to the object. The focal length of the lens is

  A   15 cm
  B   30 cm
  C   45 cm
  D   60 cm

7 Older people often suffer from a condition of the eye known as cataract. The lens of the eye becomes opaque, so that light cannot travel through it. In some cases it may become necessary to remove the lens of the affected eye. If this were to happen, think about ways in which the sight of the affected eye could be improved by the use of spectacle lenses. Draw a diagram to show how such an arrangement might work to improve the sight of the eye, and form an image on the retina. Do you think that surgeons should try to find ways of curing the condition of the eye rather than removing the lens?

8 A fish is swimming in a shallow stream of water, and is seen by an observer standing on a bank of a stream. Draw rays (two at least) from a point on the fish to show *where* the observer must aim in order to be able to spear the fish.

9 In the diagram, EF and FG are plane mirrors at right angles to each other. QO is a ray of light and OP is the normal.

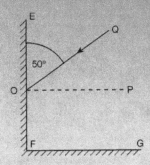

Figure 6.61

  (i)   What are the angles of reflection of the rays reflected from (a) EF, (b) FG?
  (ii) Copy the diagram and continue the ray QO to show its path after reflection from *both* mirrors.

10 (i)   Explain what is meant by the terms (a) pitch, (b) loudness, and (c) timbre.
   (ii) Explain how sounds are produced by a guitar. How would you change the pitch and loudness of the sounds produced by a guitar?
   (iii) What is the difference in the method of producing sounds in (a) a guitar, and (b) an instrument such as the clarinet or trumpet?

Investigating organisms

OBJECTIVES

- Identify and classify organisms in a habitat
- Explain the differences and interdependence between plants and animals
- Discuss the relationships between organisms
- Investigate soil as a part of the habitat

All of us have a home – a place where we live, with an address. The place where an organism lives is called its habitat. At home we have relationships and interact with other people. Organisms in a habitat also have relationships and interact with each other. As we investigate organisms, we will investigate their habitats, their relationships and their role in the habitat.

Figure 7.1 *Many different organisms live in this habitat*

7.1 Classifying organisms

In Unit 4 of Book 1, you learnt that organisms are often classified into four main groups. These are viruses, bacteria, plants and animals. In this section we are going to look in more detail at the classification of plants and animals. The plants make up the **Plant Kingdom**, while the animals make up the **Animal Kingdom**.

On your own

1 Examine an aquarium and try to observe the ways in which the fish, plants and other organisms show the signs of life. Record the data in your note book.

2 List those signs of life that are easily observable and those that take place over a long time.

3 Can you observe any differences in the way that the plants and animals in the aquarium show the signs of life?

Plants and animals can be further divided into smaller groups. We will now look at a way in which the Plant Kingdom can be classified.

ACTIVITY 7.1 TRANSPORT OF NUTRIENTS IN A PLANT

You are going to investigate one of the features used to classify plants.

You will need a beaker, some coloured ink or food colouring, a sharp knife or razor, a shoot of a plant with a clear stem (for example shining bush), and a dropping pipette.

water + red ink

Figure 7.2

1 Pour a little water into the beaker and add two or three drops of coloured ink.

2 Cut across the stem of the plant shoot and put the shoot so that its end is in the coloured solution.

3 Leave the shoot overnight, then take it out of the solution. Examine the shoot from the side. What do you see?

4 Cut across the stem and examine the cut end. What do you see?

Water and food dissolved in the water are carried up the stem of the plant in little tubes. You can see where the tubes are because the coloured solution has been carried up them.

The Plant Kingdom can be divided into two main groups, **vascular** plants and **non-vascular** plants. The vascular plants have tiny tubes to carry food and water. The non-vascular plants do not have these tubes to carry food and water.

Both the non-vascular and the vascular plants can be sub-divided into smaller groups as shown in Figure 7.3. The members of each group have similar features to other members of the same group.

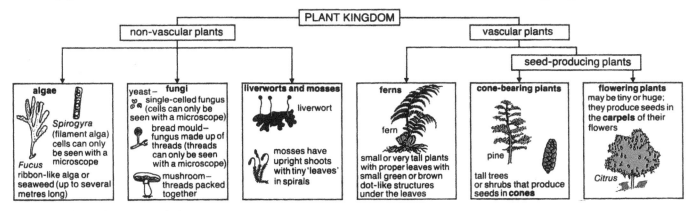

Figure 7.3 *The major groups of plants*

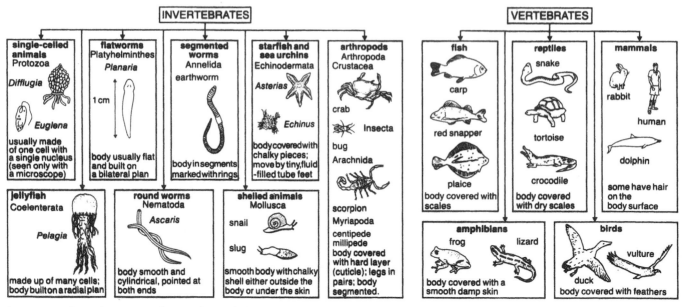

Figure 7.4 *The major groups of invertebrates (left) and vertebrates (right)*

The Animal Kingdom can be sub-divided in a number of ways. One of the most common methods is to place the animals into two groups: **invertebrates** (animals without a backbone, or vertebral column) and **vertebrates** (animals with a backbone). These groups can be further divided as shown in Figure 7.4.

Plants and animals

There are many differences between plants and animals. Here are two of them:

1   Plants show **growth movements** while animals move from one place to another – this is called **locomotion**.

2   Plants manufacture food while animals depend on plants or other animals for food.

Plants and animal cells

Plants and animals are different because the unit – the cell – that makes up the plants and animals is different.

Figure 7.5 shows a typical plant cell and a typical animal cell.

On your own

Look at the diagrams in Figure 7.5 carefully. Then make a list of the similarities and differences between a typical plant cell and a typical animal cell.

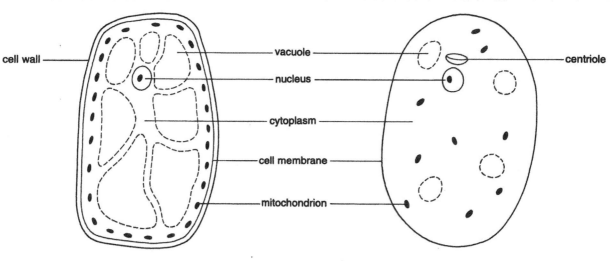

Figure 7.5   typical plant cell   typical animal cell

ACTIVITY 7.2 EXAMINING CELLS UNDER THE MICROSCOPE

You are going to examine plant and animal cells under the microscope.

You will need a light microscope, a piece of an onion bulb, a pair of tweezers, two clean microscope slides, cover slips, a sterile glass rod, and tissue stain.

1. Peel off the upper layer of the onion leaf with the tweezers and place a small piece of it in a drop of water on the slide. Place the cover slip on it.

2. Pass the sterile glass rod gently over the inside of your cheeks. Then rub it on the glass slide. Ask your teacher for some tissue stain, and place one drop on the slide over your smear. Place the cover slip on it.

3. Examine both slides under the microscope. Look at the plant cell carefully. Make a simple drawing of what you observe.

4. Examine the animal cell carefully. Make a simple drawing of what you observe.

   Do you observe all the structures in the diagram of the typical animal or plant cell? Why do you think that this is so?

7.2 Investigating habitats

A place where animals and plants live is called a **habitat**, e.g. a pond or sea shore. In each type of habitat you will find different types of plants and animals. Sometimes the animals have protective coloration (that is they have the same colour as their surroundings) or a protective shape which makes them difficult to find in their habitat. These features are examples of **adaptations**. Every habitat has certain conditions which make it suitable for some organisms but not for others. These conditions make up the **environment**.

The environment can be divided into two parts: the **physical factors** and the **biotic factors**. The physical factors include temperature, rainfall, light and humidity. The biotic factors are all the organisms in the habitat.

When investigating the environment it is necessary to look at both the physical and biotic factors.

Figure 7.6 *An example of a coastal habitat*

Measuring physical factors

As mentioned in Book 1, we measure temperature using a **thermometer**.

We measure rainfall using a **rain gauge**. You can make a simple rain gauge with a beaker and a funnel.

Figure 7.7

This is left outside in the open, away from large plants. The amount of rainwater collected in it can be measured using a small measuring cylinder. If you are comparing the rainfall in two habitats or at different times, you need to use funnels of the same size. Why?

In order to measure the amount of light present, we use a **light meter**. These are expensive, so for our investigations we will just use our eyes and say whether the environment is sunny, shady or dark.

Figure 7.8

133

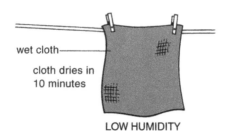

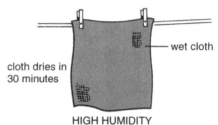

Figure 7.9

Humidity refers to the amount of water vapour in the air. If the humidity is low, there is only a small amount of water vapour in the air. The air is dry.

If humidity is high, there is a lot of water vapour in the air. The air is **humid**.

To estimate humidity, take a piece of cloth about 5 cm square. Dip it in water. Shake off the excess water and leave it in the air. Check the time it takes for the cloth to dry. If it dries quickly, the air is dry. If it takes a long time to dry, the air is humid.

On your own

1 Measure the temperature, light and humidity in the school yard. Record your observations in your note book.
Did you record the time of day?
Why is this important?

2 Using your simple rain gauge, measure the amount of rainfall in your school yard for five days.
Find out if more rain falls in the morning or in the afternoon.

You can collect plants and animals from their habitat, but make sure that you:

1 take only what you need,

2 collect them and carry them with care,

3 replace the parts of the habitat, e.g. stones and logs, as you found them,

4 clearly label (in pencil) all bags and bottles,

5 keep the animals with the plants on which they were found,

6 if possible, put the organisms back when you have finished with them.

ACTIVITY **7.3** **STUDYING A HABITAT**

You are going to use the information you have been given to make a careful study of one habitat.

You will need equipment for collecting organisms and equipment for measuring physical factors of the environment.

1 Select a habitat. It may be a rocky shore, a part of the school garden, or under a log or a rock.

item	use
	pair of scissors to cut pieces of plant
	knife or penknife to scrape off organisms from trees or rocks
	forceps to pick up small organisms
	small paint brush to pick up small organisms

item	use
	plastic bags to carry specimens
	specimen tube, tin, jar, with small holes in the lid to carry specimens
	roll of sticky-backed paper to write on and stick as labels on the containers
	plastic bowl or dish to sort organisms

Figure 7.10 *Things you need when collecting organisms*

2 Observe and record different types of plants in the habitat. Your description of each type of plant may include its height or length, the shape and size of its leaves, and the size, shape and colour of its flowers and/or fruit.

Record how **abundant** each type of plant is: plentiful (very many), sparse (not very many), or rare (very few indeed).

3 Collect a leaf from each different type of plant. Do not forget to label the leaf, noting from which plant it came.

4 Collect a sample of different animals found in the habitat. Put them in containers and label the containers, noting where the animals were found. Record how abundant each type is.

5 Measure the physical factors in the habitat you have selected.

6 In the laboratory or classroom, draw each type of leaf you have collected. Note any differences in size, shape or texture.

7 Examine carefully the animals you have collected using a hand lens or, if necessary, a microscope. Make simple drawings of each animal. Make some notes about its size and appearance.

How is it adapted to its habitat?
Does it move?
How does it feed?
Using the classification scheme on page 132, in which group would you place the animal?

8 Prepare a report of your investigation. Indicate the location and size of the habitat you selected.

Preserving plant and animal specimens

If it is not possible to return the organisms to their habitat when you have finished with them, you may wish to preserve them. You could have a display with your report.

Place the plant specimen between the pages of a small newspaper and place a heavy weight on the paper. Make sure that the specimens are not folded.

You may preserve insects and other arthropods by mounting them individually on a pin, then display them on a piece of cardboard, with appropriate labels.

Other animals may be preserved in liquids such as alcohol or formalin. Ask your teacher for assistance. Make sure the containers are securely covered and clearly labelled. Be very careful not to spill any formalin or let it touch your skin. If it does, wash it off immediately with plenty of water.

ACTIVITY **7.4** **COMPARING HABITATS**

You are going to use the same methods as in the last activity to compare two habitats.

1 Select two different habitats.

2 Compare the physical factors in the two habitats.

Figure 7.11a   *Caroni swamp in Trinidad*

Figure 7.11b   *Kalahari desert in southern Africa*

What are the main differences between these two habitats?

135

3 Observe the plants and animals present.
Do the two habitats have any plants or animals in common?

4 Make a list of any features of the plants and animals that seem to be adaptations to the particular habitat in which they are found. Give reasons for the features you select.

5 Using the approach suggested in Activity 7.3, write a report showing the main differences between the two habitats.

Organisms of the same kind, or **species**, have features in common and are able to mate with one another.

The total number of organisms of the same kind that live in a particular habitat at the same time form a **population**. These organisms may vary in age. For example, all the ants living in a nest in the soil form a population.

All plant and animal populations living in a particular habitat make up a **community**. In a pool, all the different populations of fish, water weeds, tadpoles and so on, make up a community. A community therefore contains different species of plants and animals. The different species of plants and animals interact with each other. For example, some organisms may be eaten by others. The plants and animals also depend on the physical factors in the environment for their survival.

On your own

1 How many species of plants and animals did you find in the habitats you studied?

2 List any ways in which the plants and animals in the habitats you studied were dependent on each other.

ACTIVITY 7.5 OBSERVING A HABITAT OVER TIME

You are going to observe a habitat over time to see what changes take place.

1 Select a habitat which you can study over a period of three to four weeks.

2 Choose convenient times in the morning and afternoon and observe the habitat at these times every day.

3 At each observation time, measure the physical factors.

4 Record the approximate number of different species of plants present and then note any increase in size, number of branches, appearance of flowers or replacement of flowers with fruits.

5 Observe the animals, noting any changes in behaviour or abundance at different times of the day.

6 Write a report of your investigation.

7.3 Identifying organisms

There are so many different kinds of living organisms in the world, that it would be impossible to learn to recognise them all.

If we find an organism that we need to identify, we use a key. The simplest type of key to use is a **branching key**.

We have a list of things – dog, cat, bird, snake, rock, mango, telephone, tree.

To find the name of any of the objects or organisms, first we examine it closely, then we go along the branch which describes it. We choose the next suitable descrip-

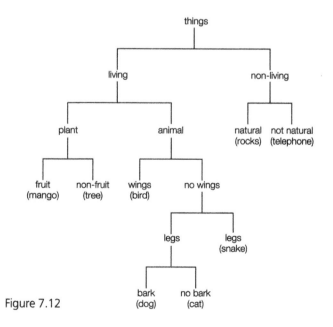

Figure 7.12

we come to the **name** that identifies the object or organism.

The problem with branching keys is that they take up a lot of space. We can convert branching keys into **numbered keys**.

	Living	(1)
	Non-living	(6)
1	Plant	(2)
	Animal	(3)
2	Fruit	Mango
	Non-fruit	Tree
3	Wings	Bird
	No wings	(4)
4	Legs	(5)
	No legs	Snake
5	Bark	Dog
	No bark	Cat
6	Natural	Rock
	Not natural	Telephone

Down the left of this key is a row of numbers. Each number has two descriptions. Choose one. Opposite each description is the next number to go to; for example, if the object is non-living you go to 6; if it is natural then it is a rock, if it is not natural then it is the telephone.

Making your own key

The first step in making a key is to split the organisms to be included in the key into two groups, for example living and non-living.

Each group is then split into two smaller groups.

We continue doing this until there is only one organism left in a group. We then include its name.

For a branching key, the description which fits the members of each new group is put at the end of a new branch.

On your own

Collect a sample of different kinds of peas and beans. When you have about ten or more different kinds, make a branching and then a numbered key to identify them. These keys are also called **binary** and **serial**. Which name do you think is used for the branching key and which for the numbered? Give reasons.

7.4 Animals in a habitat

When you compared habitats or studied a habitat over a period of time, did you notice that the animals fed on different kinds of food?

Figure 7.13

Some animals feed only on plants. They are called **herbivores**. Some animals feed only on other animals. They are called **carnivores**. Animals that feed on both plants and animals are called **omnivores**.

On your own

From the list of animals you observed in the habitats you have studied, classify them using the type of food they eat.

Make a list of any differences you could observe in their feeding habits and mouth parts.

Relationships between organisms

As you observed the different habitats, did you notice that organisms (plants and animals) form relationships of one kind or another with each other?

When an animal kills and feeds on another animal, they are said to have a **predator/prey relationship**. The predator kills and feeds on the prey. The predator and prey populations balance each other.

On your own

Identify a predator/prey relationship in the habitat you have studied.

What do you think would happen if there was a sudden increase in the prey population? Indicate on a multi-line graph what would happen to each population over time.

Other relationships

Many organisms are found living together in unusually close and special relationships. The relationships that take place between different species are called **symbiosis**.

There are many kinds of **symbiotic** relationships, some closer than others. They fall into three categories: **mutualism**, **commensalism** and **parasitism**.

Mutualism is a relationship between two species from which both partners get some benefit. They may be unable to live separately. **Lichens** grow on bare rocks, the stems of large plants and other places. Lichens consist of two organisms – an **alga** and a **fungus** – joined closely together. The alga makes food for the fungus by photosynthesis while the fungus gives the alga moisture and anchorage.

Figure 7.14a   *A lichen living on rock*

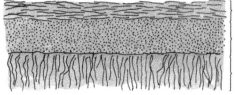

top fungal layer
(holds water)

green algal layer
(makes food)

lower fungal layer
(anchorage)

Figure 7.14b   *Cross-section of a lichen*

The root nodules on the roots of some plants, e.g. peas and beans, contain bacteria which live in a state of mutualism with the plant. The bacterium called **Rhizobium** takes nitrogen from the air and changes it into a form that the plant can use. In return the bacterium gets food from the plant.

Commensalism is a relationship between two species in which one partner, the **commensal**, gets some benefit; while the other, the host, neither gains or loses.

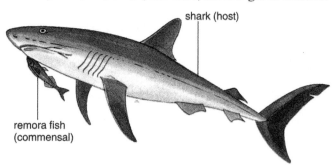

shark (host)

remora fish
(commensal)

The **remora** is a small fish which attaches itself to a shark by a sucker on its head. It gets carried around by the shark with its mouth free to catch bits of food which the shark may drop. The shark appears to be completely unaffected by this.

Parasitism is a relationship between two species in which one, called the **parasite**, gets food and shelter from the other, called the **host**, while causing some harm to it. The parasite may eventually kill the host.

Tapeworms live in the intestines of people who have eaten infected meat. Their long, flat bodies, up to 6 metres in length, absorb digested food from the host. Special chemicals keep the worms from being digested.

7.5 Interdependence between plants and animals

In any habitat, plants and animals are interdependent. Each contributes to the survival of the other. The plants provide food, protection and sites for building nests for animals. The animals, on the other hand, carry out pollination to enable reproduction of the flowers. They are also involved in fruit and seed dispersal.

A habitat supports a number of different species of plants and animals. All the plants and animals found in a particular habitat are known as a **community**.

Food chains

In Book 1 we discussed the flow of energy through a community. This energy flow was illustrated in a food chain, e.g.

grass ⟶ cow ⟶ man

The green plants – **the producers** – manufacture food by photosynthesis using energy from the Sun. When the green plant is eaten by the herbivore – **primary** consumer – the herbivores are eaten by carnivores – the **secondary consumers**. In some food chains small carnivores are eaten by larger ones. These are called **tertiary consumers**. In every habitat there are food chains.

On your own

plant ——→ caterpillar ——→ bird ——→ cat

In the food chain above, identify the producer and the various consumers. Explain the flow of energy in the food chain.

Food webs

Certain organisms in a habitat get their food without killing. They feed on the remains of other organisms. These are the **scavengers** and **decomposers**. Scavengers feed on the flesh of dead animals or on dead bits of plants. Some feed on animal droppings. They assist in keeping the environment clean. They may also be eaten by carnivores.

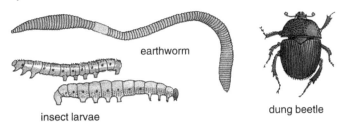

Figure 7.16 *These animals are all scavengers*

Decomposers are the fungi and bacteria which cause dead things to decay. When they are finished all that is left are inorganic minerals which return to the soil. In Book 1 we looked at the nitrogen cycle which contained decomposers. Decomposers do an important job in recycling minerals which plants need for healthy growth.

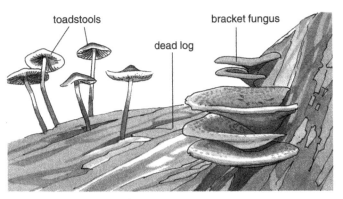

Figure 7.17 *Fungi are decomposers*

Most organisms in a habitat feed on more than one kind of food. This means that they must belong to more than one food chain.

We can look at all the food chains in a habitat. If we join them together we get a complex picture called a **food web** (see Figure 7.18). This shows a greater number of energy relationships than a food chain.

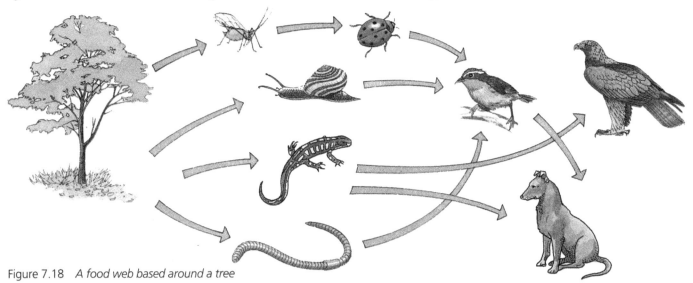

Figure 7.18 *A food web based around a tree*

7.6 Energy and food relationships

Do you remember that plants make carbohydrates using energy from the Sun? Animals eat the food made by plants, either directly by eating plants or by eating other animals that eat plants. In this way, animals can get the energy they need for living.

For example, in a certain habitat leaves may be eaten by grasshoppers. Grasshoppers are eaten by lizards and lizards are eaten by birds. This chain can be shown in a diagram as shown on the left below.

You may have found that in the habitats you have studied, the food–energy relationships are more complicated than this because animals eat more than one type of food.

Using our simple example, we can also show the food–energy relationship in another way as shown on the right below.

Do you notice that, going from the top to the base of the pyramid, there is a greater number of organisms at each level? Can you suggest why?

The amount of food and energy an organism needs depends on its body size and on how active it is. Grasshoppers, for example, require a lot of energy for movement. An enormous number of leaves are needed to feed grasshoppers. Very many grasshoppers will feed a few lizards. And a few lizards are needed to feed each bird. As you go from the base of the pyramid (the **food producers**) to the top (the **final consumers**) the body size of the organisms increases and their numbers decrease.

As you go from the base of the pyramid to the top, the total energy stored in organisms at one level also decreases. For example, grasshoppers obtain the energy they need from the leaves they eat. They use some of this energy for moving about and other activities, and they store some of the energy in their bodies. When lizards eat the grasshoppers, they obtain the energy that the grasshoppers have stored in their bodies. They do not obtain the energy which the grasshoppers have already used up.

Figure 7.20   *A grasshopper*

If there is a large number of feeding levels in a pyramid, a lot of energy is used up by organisms in the pyramid. The smaller the number of feeding levels, the more energy is available for the final consumer. In practice, this limits the number of feeding levels in a food pyramid and the number of final consumers that can be supported by the plants in a given area of the environment.

If humans wanted to get the greatest amount of energy from a given area of the environment, at what level in this food pyramid should they feed?

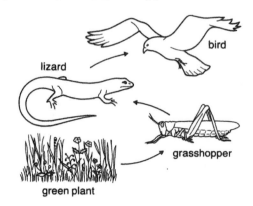

Figure 7.19

possible number of each
type of organism on an
area of uncultivated land

large birds	5–10
lizards	70–100
grasshoppers	many
green plants	very many

7.7 Investigating soil

One factor of the environment that we have not mentioned so far is the soil. Have you ever heard farmers saying that they can't grow certain crops because of the soil in their fields? Different countries in the Caribbean often grow different crops, although many factors in the environment seem similar. This is partly because they have different soils. The soil is a physical factor which has an influence on what organisms can survive in a habitat; plants are especially dependent on the nature of the soil.

Most plants get their water and minerals from the soil. Many properties of the soil determine the amount of water that can be taken up by a plant. These include the composition of the soil, the upward movement of water in the soil, and how much water and air there are in the soil. The acidity or alkalinity of the soil has an important effect on which plants will thrive in it.

The diagrams on the right and on page 142 suggest investigations you can do on the properties of soil.

Figure 7.21

Professor Nazeer Ahmad is the Head of Department of Soil Science in the Faculty of Agriculture at the University of the West Indies in Trinidad.

For the last 25 years, Professor Ahmad has done extensive research on the basic properties, behaviour and management of soils of the Caribbean. He has developed a classification of Caribbean soils according to appropriate land capability and uses. He has also done extensive work on soil management for pasture and rice production. Recently, he has researched how nitrogen fertilizers may be used to greatest advantage.

Professor Ahmad is an internationally known soil scientist and has published numerous articles in international science journals and books and presented many papers at international conferences.

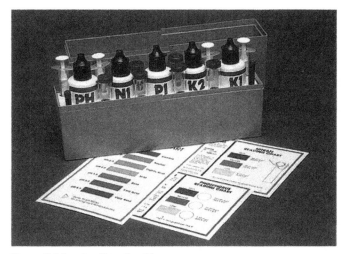

Figure 7.22 *A soil testing kit*

ACTIVITY 7.6 INVESTIGATING PROPERTIES OF SOIL

You are going to find out about the soil in your school garden.

You will need some soil, test tubes, water, glass tubes, a crucible, an oven, scales, measuring cylinders and pH paper.

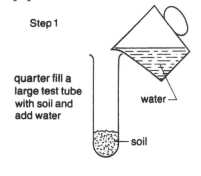

Step 1

quarter fill a large test tube with soil and add water

water

soil

Step 2

shake well then allow the soil to settle

humus (floating)
small clay particles (in suspension)
large clay particles
silt
sand
gravel (small stones)

Figure 7.23 *Experiment to test the composition of soil*

Test the composition of the soil in your school garden. Does this soil contain sand or clay or both? What other constituents does it contain?

1 Upward movement of water through soil:

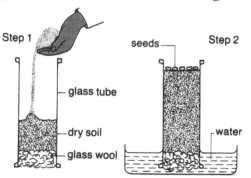

Note the time as you place the tube in the water. Observe the tube at intervals and record what you observe. Record the time taken for the seeds to start germinating.

2 Water in soil:

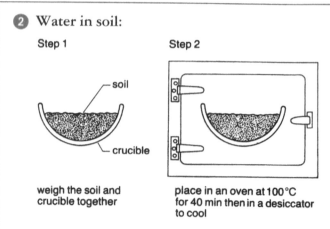

Re-weigh the soil and crucible. Repeat step 2 and then re-weigh. Continue until there is no change in mass.

3 Air in soil

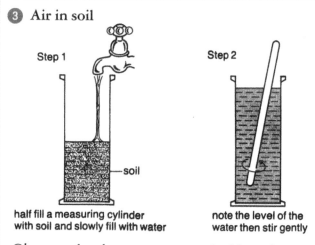

Observe what happens as you stir. Note the new level of the water. What do you deduce?

4 pH of the soil

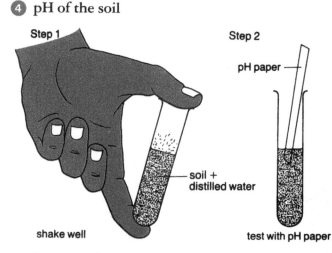

Find the pH of soil in the school garden.

Figure 7.24   *Four soil experiments*

On your own

Find the answers to the following questions from books, by asking farmers, and/or by carrying out experiments.

1 Which plants grow well in acid soils and which plants grow well in alkaline soils?

2 What can farmers do to change the acidity of their soil?

3 What are the problems of farming on soil that is very sandy?

4 What are the problems of farming on soil that is very clayey?

5 What can farmers do to alter the water drainage in their soil?

6 The soil is a physical factor which affects living organisms. But the soil is, in turn, affected by the organisms that live in and on it. In what ways can living organisms affect the soil?

7.6 Food production in plants

Investigating leaves

You have learnt about the importance of seeds, fruits and flowers to a plant. Do you know why leaves are important to plants? The following activities will help us to understand the importance of leaves.

 ACTIVITY **7.7** **FINDING OUT ABOUT LEAVES**

You are going to classify and examine leaves.

You will need leaves from about 20 different plants, graph paper and ruler, a scalpel, a microscope with slides and cover slips.

1. Collect 15–20 leaves from different plants. Try to get as many different leaves as possible. Classify them, making a note of your reasons for putting them into the different groups. Make a numbered key for the leaves.

2. Examine a leaf. Make a simple drawing noting the names of the parts. Can you find all the parts labelled in the diagram below?

   Is there any difference in the appearance of the upper and lower surfaces? Suggest reasons for any differences.

   Examine the arrangement of the veins in different leaves. What sort of pattern do they form?

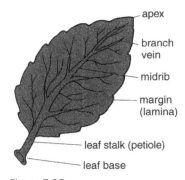

Figure 7.25

3. Take one of the leaves. Place it flat on squared (graph) paper and trace around it in pencil. From the number of squares covered by the leaf, work out the approximate surface area of the leaf. (You will need to find the area of the squares on your graph paper using a ruler.)

What would be the total surface area of leaves if the plant had 50, 100 or 200 leaves like the one you measured?

Why does a plant need a large leaf area?

4. With a scalpel or sharp razor blade, cut very thin slices of a leaf. These slices are called **transverse sections** or **cross sections**.

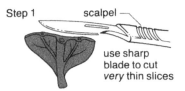
Step 1 scalpel
use sharp blade to cut *very* thin slices

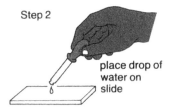
Step 2
place drop of water on slide

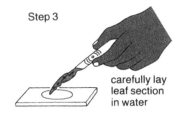
Step 3
carefully lay leaf section in water

Step 4
cover with cover slip

cover slip
water
leaf section
slide
view from the side of completed slide

Figure 7.26

Place a drop of water on a microscope slide. Carefully place the leaf section in the drop of water on the slide. Cover the section with a cover slip. Examine the section under a microscope.

Did you identify the structures shown in Figure 7.26?

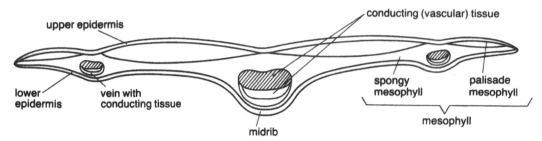

Figure 7.27 *Cross-section of a leaf*

The surface of the leaf is the **epidermis**. The upper epidermis is covered by a waxy later called the **cuticle**. In the lower epidermis there are small openings called **stomata**. These allow gases to pass into and out of the leaf.

Leaves and photosynthesis

The start of every food chain is a plant. Why is this so?

Plants produce the food on which all life depends by the process of photosynthesis. They produce simple sugars which they convert into starch and store. In this section we are going to investigate some of the conditions required for photosynthesis to take place.

The activities you will carry out require you to test leaves for starch. So first, carry out the following activity to check that you understand how to do this test.

ACTIVITY **7.8** **TESTING A LEAF FOR STARCH**

You are going to learn the procedure for testing a leaf for starch.

You will need a beaker, water, 70 per cent ethanol, a test tube, cotton wool, a dropping pipette, a watch glass, a Bunsen burner and a leaf.

① Boil the leaf in water for one minute. This bursts any grains of starch present.

② Place the leaf in a test tube which is one-third full of ethanol. Put a plug of cotton wool loosely in the mouth of the test tube. Place the test tube in a beaker of water. Warm the water gently until the ethanol simmers, *but do not boil the water*. The leaf gradually loses its green colour. This allows the test results to be seen clearly.

Why does the leaf lose its green colour?

③ Dip the leaf in boiling water. The leaf will have become hard in the ethanol. The boiling water softens it again, so that the iodine can enter it.

④ Soak the leaf in iodine solution for two minutes. Then hold the leaf up to the light. If starch is present in the leaf, the iodine will have turned blue-black in colour.

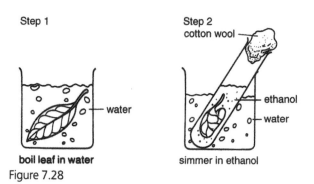

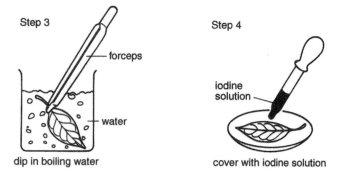

Figure 7.28

INVESTIGATIONS ABOUT PHOTOSYNTHESIS

You are going to design and carry out activities to investigate the following hypotheses:

❶ Light is necessary for photosynthesis.

❷ Chlorophyll is necessary for photosynthesis.

❸ Carbon dioxide is necessary for photosynthesis.

❹ The gas given off during photosynthesis is oxygen.
Please show your teacher your design before you start the activity.
For these activities you will need healthy potted plants.
Before you put them in the test conditions, you need plants that have no starch in them. Why should this be so?
To remove the starch from potted plants, keep the plants in the dark for at least two days. This ensures that all the starch present has been removed. We say that the plants have been **destarched**.

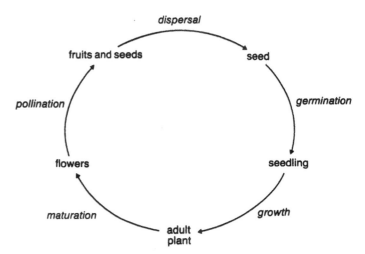

Figure 7.29   *The life cycle of a flowering plant*

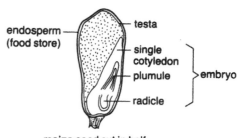

maize seed cut in half

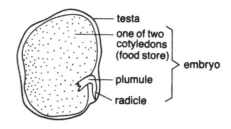

bean seed cut in half

Figure 7.30   *A maize seed has one cotyledon; a bean seed has two*

 ## 7.9 The seeds of flowering plants

In Unit 6 of Book 1 you learnt that flowering plants produce seeds. Let us examine some seeds and see what part they play in the life cycle of a plant.

A seed has two main parts:

1   an outer covering, the **seed coat** or **testa**,

2   the **embryo** or young plant.

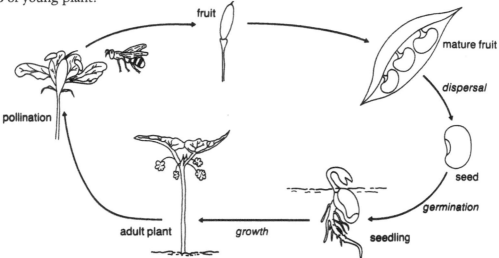

Figure 7.31

The embryo is made up of the **radicle**, which develops into the root; the **plumule**, which develops into the shoot; and one or two **cotyledons** or seed leaves, which may store food. Some seeds, such as maize, store food in the **endosperm** rather than the cotyledons.

Seeds germinate to produce **seedlings** or young plants. The young plants grow into adult plants. The adult plants produce flowers for sexual reproduction. Pollen is transferred from the anthers of a flower to the stigma, usually of another flower, by insects, birds or the wind. After pollination, the flower gives rise to fruits which contain seeds. The fruits or seeds are dispersed. If they fall on a suitable habitat, the seeds germinate to give seedlings, and the cycle is repeated.

On your own

1. Collect a maize seed, a bean seed and any other types of seed you can find. In your note book draw and describe the outside of the seed.

2. Cut open each seed, and describe the parts you observe.

ACTIVITY 7.10 GERMINATION OF SEEDS

You are going to investigate the conditions necessary for the germination of a bean seed.

You will need four beakers, blotting paper, water, bean seeds, oil, fine soil.

1. Line the four beakers with blotting paper; put seeds between the blotting paper and the side of the beaker; and put soil in the beakers to hold the blotting paper in place.

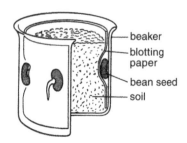

Figure 7.32

2. Control the conditions in the four beakers as shown in Figure 7.33 and observe them every day. In which beaker(s) do the seeds germinate?

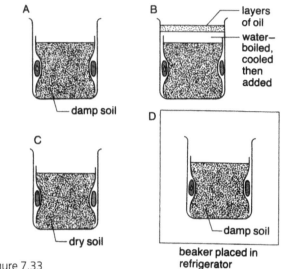

Figure 7.33

Why was the water first boiled and then cooled in beaker B?
What conclusions do you draw from your observations?

ACTIVITY 7.11 OBSERVING SEEDLINGS

You are going to observe how the parts of a seed develop.

You will need some large seeds, pots with damp soil.

1. Cut open some of the seeds and draw the parts.

2. Plant the seeds and leave them to germinate.

3. As the seedlings develop, take one out of its pot every few days, and draw the parts.

Do the cotyledons come up above the ground?
What provides food for the seedling before the first leaves appear?
What provides food for the seedling once leaves have appeared?

Different seedlings show slightly different stages of development. This is shown in Figure 7.34.

Seed dispersal

What do you think would happen if all seeds fell immediately beside the parent plant? Would all the seedlings

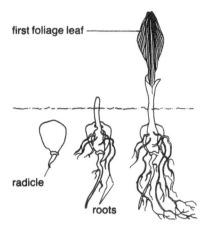

first foliage leaf

radicle

roots

stages in germination of maize seed

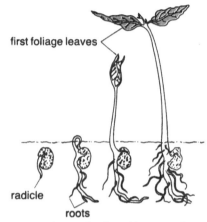

first foliage leaves

radicle

roots

stages in germination of bean seed

Figure 7.34

get the light they need to develop? Would they get all the minerals they need from the soil?

It is an advantage for seeds to be **dispersed**. If all seeds fell by the parent plant, they would not thrive. When seeds are dispersed, many fall on unsuitable ground and fail to germinate. However, some are carried to a suitable habitat where they can develop into a new plant.

Seeds may be dispersed in various ways:

by animals, e.g. clinging seeds called burs catch onto the coats of animals and onto clothes; brightly coloured berries are eaten by birds and seeds inside them are dispersed in their droppings.

by the wind, e.g. cotton seeds have fine threads attached which are easily blown by the wind.

by explosion of the fruit, e.g. the fruits of the sand box tree explode as they dry, hurling the seeds away from them.

by water, e.g. the coconut fruit has pockets of air in its fibres and floats on the sea which carries it to distant shores.

by humans, e.g. people plant crop and flower seeds in places where they are likely to survive.

Figure 7.35 *Suggest the method of seed dispersal of each of these three plants*

The red mangrove, found around the coast in many parts of the Caribbean, is unlike most other trees in the way its seeds develop. The seed grows while it is still attached to the tree. The seedling grows on the seaward side of the tree. When it is about 18–36 cm long, it drops from the tree and penetrates into the mud where it develops into a new tree.

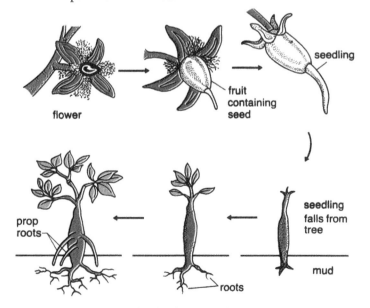

flower

fruit containing seed

seedling

seedling falls from tree

mud

prop roots

roots

Figure 7.36 *Stages in the development of the red mangrove tree*

147

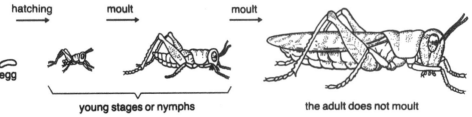

hatching → moult → moult

egg

young stages or nymphs

the adult does not moult

the number of moults varies from insect to insect

Figure 7.37

7.10 Life cycles of insects

Two different types of life cycle are shown by insects.

Some insects, like grasshoppers and cockroaches, lay fertilized eggs in which a zygote develops.

When the zygote is fully developed, the shell splits and the young insect emerges from the egg.

The young grasshopper or cockroach is just like the adult except that it is smaller, has no wings and cannot reproduce. The young is called a **nymph**. It sheds its skin or **moults** several times (see Figure 7.37). Each time it sheds its skin, it increases in size before the new skin hardens. The wings also develop. After the final moult, the wings are fully grown. The insect's reproductive organs are now fully developed and it is able to reproduce.

This pattern of development is called **incomplete** or **direct metamorphosis**. ('Metamorphosis' means 'change in structure'.)

In other insects, such as butterflies, houseflies or mosquitoes, the young that hatches from the egg is unlike the adult. It is a worm-like **larva**. The larva is either a caterpillar, a grub or a maggot.

The larva is the feeding stage which moults and grows several times. When it is fully grown, it changes to the next stage: the **pupa** or **chrysalis**. The pupa is usually stationary. A lot of reorganization takes place in the pupa. After some time, the adult insect emerges from the pupa.

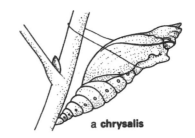

Figure 7.39

a **chrysalis**

This pattern of development is called **complete** or **indirect metamorphosis**. Complete metamorphosis has the advantage that the larva and the adult do not compete with each other for food. They eat different types of food at the two stages.

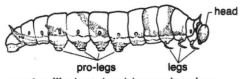

head

pro-legs   legs

a **caterpillar** has a head, legs and pro-legs

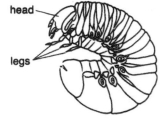

head

legs

a **grub** has a head and legs

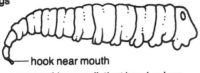

hook near mouth

a **maggot** has no distinct head or legs

Figure 7.38

On your own

1. Collect insect eggs, nymphs or larvae.

2. Keep them in an appropriate place where they have enough of the right type of food, the temperature is similar to their natural habitat, and they can get air to breathe.

3. Observe their stages of development. Keep records of the size, the number of moults, and the length of time taken for each stage of their development.

4. Observe what foods they prefer at different stages.

5. Make simple drawings of the stages.

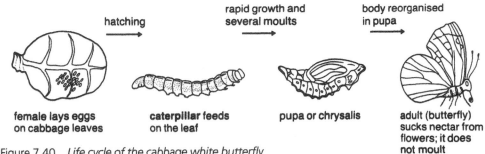

female lays eggs on cabbage leaves → hatching → caterpillar feeds on the leaf → rapid growth and several moults → pupa or chrysalis → body reorganised in pupa → adult (butterfly) sucks nectar from flowers; it does not moult

Figure 7.40   *Life cycle of the cabbage white butterfly*

Figure 7.41   *Caterpillar, pupa and butterfly*

Summary

Here are some of the main ideas you learnt in this unit.

- The Plant and Animal Kingdoms can be sub-divided into smaller groups.

Figure 7.42   *Pupils helping to protect young turtles from predators*

- There are a number of differences between plants and animals and plant and animal cells.

- Habitats differ in both physical and biotic factors.

- Organisms can be identified by means of a key.

- Animals can be classified by the type of food they eat – herbivores, carnivores and omnivores.

- Organisms in a habitat have a relationship with each other – some of these are predator/prey (see Figure 7.42); mutualism; commensalism and parasitism.

- A food web illustrates the complex energy relationships between organisms in the habitat.

- The special properties of soil enable organisms to live and grow in it.

- Leaves are important to plants – they manufacture food.

- Seeds play an important part in the life cycle of plants. They contain the embryo or young plant.

- Insects have two different types of life cycles – incomplete or direct metamorphosis and complete or indirect metamorphosis.

QUESTIONS

1 What materials from the environment are used by producers?

    A   light, water, CO_2
    B   light, soil, air
    C   water, air, soil
    D   water, soil, CO_2

2 Which of the following is a difference between a herbivore and a carnivore?

    A   A herbivore eats a consumer while a carnivore eats a producer
    B   A herbivore eats all types of food while a carnivore eats meat only
    C   A herbivore eats a producer while a carnivore eats a consumer
    D   A herbivore eats meat only while a carnivore eats plants only

Below is a food web. Questions 3–6 relate to this diagram.

3 Put in the arrows in the food web – there should be at least twelve arrows in all.

4 From the diagram, list two secondary consumers.

5 From the diagram, draw and label a food chain with a producer, and a primary, secondary and tertiary consumer.

6 From the diagram, identify three predator/prey relationships.

7 Snake, mouse, hawk, pigeon, spider, fly, leaf, cherry. Using this list, make a binary and serial key.

8 Using the list of organisms provided below, construct a food web. Circle each producer and underline each consumer. Place a (*P*) next to the organism for primary consumer, (*S*) for secondary consumer, and (*T*) for tertiary consumer.

    corn decomposer deer dog grass mouse owl iguana seed-eating bird snake lizard tree

9 Some animals can be both primary and secondary consumers. Describe how humans can be both primary and secondary consumers.

10 What is meant by life cycle?
With the aid of diagrams, compare the life cycle of a flowering plant and a named insect.

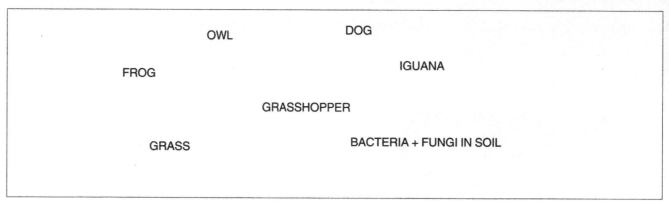

Figure 7.43

Investigating our marine environment

OBJECTIVES

- Discuss the factors causing tides
- Explain the properties of waves
- Explain the effect of waves and currents in the sea
- Discuss the nature and importance of the sea

Almost everywhere in the Caribbean, we are close to the sea. It is an environment that has influenced the history, economy and way of life of the region. The sea has also influenced the plants and animals that live here. Tourists come many miles to bask in the sunshine and bathe in the clear blue Caribbean waters. The marine environment is important to us. How can we use it and preserve it?

Figure 8.1 *The Caribbean region has many beautiful beaches*

Movements of the sea

Tides

Some of your friends went to the beach and on their return they told you this story: 'We arrived at the beach early in the morning, chose a shady spot a short distance away from the water and put down our picnic basket and towels. Quickly we changed our clothes and ran into the water. We had a good time in the water, swimming and playing games, and the time passed quickly. Suddenly someone shouted "Look!" and pointed to the beach. There we saw our picnic basket just beginning to float in the water and our towels wet beside it.'

Why did the basket get wet after it had been placed away from the water?

Your friends forgot about **tides**; that is, the rise and fall of the water level of the sea at different times of the day.

When your friends arrived at the beach the water level was at its lowest point. When the sea is at this point we say it is **low tide**. After low tide, the water level gradually rises to the high or **flood level**. When the sea is at the flood level, we say it is **high tide**. After high tide, the level starts to fall again until low tide. This repetitive rise and fall of the water level is called the **flow** (rise) and **ebb** (fall) of the tide.

On your own

If you live near the sea, go down to the beach. How can you tell where the flood level of the most recent tide was? Why is it more difficult to find where the most recent low tide was?

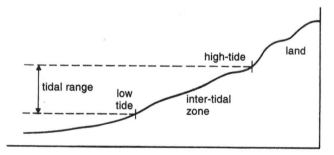

Figure 8.2 *The tidal range*

The difference in the height between the water level at low tide and at high tide is called the **tidal range**. The tidal range varies in different places, and is affected by the shape of the coast and the sea floor. In some parts of the world the tidal range may be as much as 16 m.

The area on the beach between the water level at high tide and the water level at low tide is called the **inter-tidal zone**. What factors will affect the size of the inter-tidal zone?

On your own

1. The next time you go to the sea, record the name of the beach and record the width of the inter-tidal zone. See whether the width of the inter-tidal zone is the same for all beaches in your territory at all times of the year.

2. If you live near to the sea, record the time of low tide and high tide for a period of four days.

   If you do not live by the sea or it is not convenient to record the time (you may be in school), look at the weather section of the daily newspaper for the times of high tide and low tide for a period of four days. Also record the time the Moon rises each day. You may also listen to the local weather forecast on the radio or television.

   How many times is there a high tide in each day?

   What is the length of time between high and low tide?

   What is the length of time between one high tide and the next? Is it the same as the time between one low tide and the next?

   Do the tides occur at the same time each day?

   If high tide was at 6.00 a.m. yesterday, would you expect it to be earlier or later today?

   Does the moon rise at the same time each day?

   What is the difference in the time the Moon rises on one day and the next?

   Is there a relationship between the difference in time that the Moon rises each day and the time between high tides?

Sandy beaches are not the only kind of shore in the Caribbean. Mangrove swamps develop in places where the waves are very small and the sea bed slopes gently. Sea grass beds may grow where coral fragments are mixed with sand. There are also rocky beaches with boulders or cliffs. In each of these habitats you will find a community of plants and animals that have adapted to it.

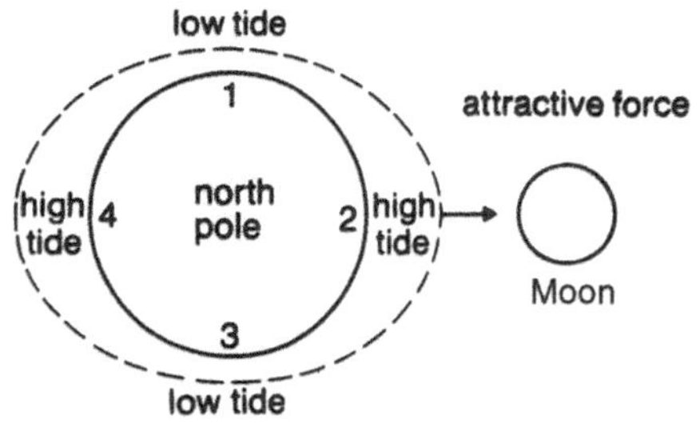

What causes tides?

When you answered the questions on page 152, you should have noticed two things:

1 the rise and fall of tides follow a regular timetable

2 the timetable of the tides resembles the timetable of the Moon.

This suggests that in some way the tides are associated with the Moon.

Remember you learnt that the Earth rotates around the Sun and the Moon rotates around the Earth. The Sun, the Earth and the Moon have a gravitational pull, and it is the gravitational pull of the Moon and the Sun that causes tides.

If there were no gravitational pull of the Moon and the Sun, the water of the seas would be evenly distributed all around the Earth and its level would not change.

Although the Sun has a greater mass than the Moon, the Moon is much closer to the Earth. Therefore, the Moon has a greater influence in causing tides than the Sun.

The Moon pulls on all parts of the Earth and the seas. Look at Figure 8.3. When the Moon is opposite point 2, the water at that point is pulled more towards the Moon than the solid Earth beneath; therefore there is high tide at point 2. At point 4 the solid Earth is pulled more towards the Moon than the water is. So a bulge of water appears at point 4; there is a high tide here too. At points 1 and 3 the level of the water falls; there is a low tide at these points.

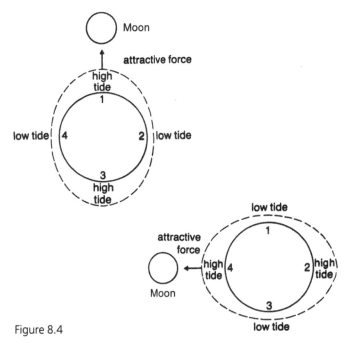

Figure 8.3

As the Moon moves round the Earth, the high tide moves round. Look at Figure 8.4. After about six hours, the Moon is opposite point 1 which, together with point 3, has a high tide. Points 2 and 4 now have low tides.

Figure 8.4

Approximately six hours after that the Moon is opposite point 4. Then there is high tide at points 4 and 2 and low tides at 1 and 3.

The Sun also affects the tides. When the Sun and the Moon are in a direct line they have a combined gravitational pull on the Earth as shown in Figure 8.5. At this time the high tides are higher and the low tides are lower than usual. These tides are called **spring tides**. Spring tides occur twice each month.

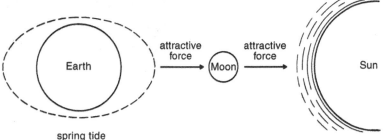

Figure 8.5 *Spring tides*

spring tide

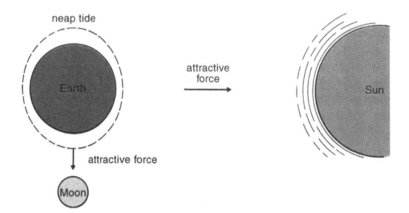

Figure 8.6 *Neap tides*

When the Moon and the Sun are at right angles to each other, their gravitational forces oppose each other. They therefore have less effect on the water around the earth. Now the high tides are not as high and the low tides are not as low as usual. The high tide is at its lowest and the low tide is at its highest. These tides are called **neap tides**. Neap tides occur twice per month.

On your own

① From your knowledge of the phases of the Moon, work out at which phases of the Moon there are spring tides and at which phases there are neap tides. Study the weather section of a newspaper over a period of time to see if you are correct.

② If possible, visit the sea shore during spring tide and neap tide and measure the tidal range. The easiest way to measure the tidal range is to find a rock on which you can mark the high tide level and the low tide level and measure the vertical height between them.

Figure 8.7

Effects of tides

Many activities, for example shipping and ocean-boating sports, are timed according to the tides. Tugs that pull huge cargo boats and passenger liners out of harbours into the deep ocean use tide tables to determine the best time to work.

Very often fishermen time their catch to follow the tides, since certain fish are brought in with the tides.

Plants and animals living on the sea shore are affected by tides. If they live on the shore within the inter-tidal zone they will be covered by water at high tide. At low tide, they will be exposed to the air.

The plants that live in this zone have to be able to withstand drying out when they are exposed to the sun and air.

Most of the animals that live in the inter-tidal zone burrow into the sand during low tide. Some small shellfish called limpets feed only when they are covered by water, and at this time they can move about on the rocks where they live. When the tide is out, they remain tightly attached to the rocks. They are so firmly attached that sea birds cannot pull them off to eat them. To prevent the sun from drying them out, they store a little sea water inside their shells.

Figure 8.8 *Limpets on rock*

If we are not careful, at high tide the water may wash away things we leave on the beach.

The tidal range in most of the Caribbean is much smaller than in many parts of the world. It is often only about 20 cm. This means that there is only a small zone which is covered and uncovered by water every day. However, many plants and animals have adapted to this environment, especially in mangrove swamps.

Waves

The movement of the sea with which we are most familiar is **waves**. Waves are produced when the sea is disturbed, usually by wind blowing over its surface.

ACTIVITY 8.1 INVESTIGATING WAVES

You are going to use a simple model to show what causes waves and what affects their size and speed.

You will need a basin of water and a three-speed electric fan.

1 Place the basin of water on the table.

2 Blow air from the fan over the surface of the water, first at slow speed, then medium, then fast. Observe what happens.
What did you notice about the *size* of the wave as the speed of the air increases?
How would you describe the *speed* of the wave in relation to the speed of the wind?
Does the wave move at the same speed, faster or slower than the air?

3 What do you think would be the effect of using a smaller basin or a larger basin? Formulate a hypothesis relating the size of the wave and the size of the container. Design and carry out a simple experiment to test your hypothesis.

In general, all waves, be they sound waves, light waves, water waves or other kinds of waves, follow the same rules.

Touch the surface of still water with your finger. You will see circles spreading out from the spot you touched. These are **ripples**, tiny waves on the surface. As these circular waves move outwards, the water surface moves up, down, up, down, forming a smooth curve. Each part of the water surface hands the motion of the wave on to the next. This is called **wave-propagation**.

If you keep on moving your finger up and down, you send out one circular wave after another. Your finger provides the energy to disturb the water surface. It is the **wave source**.

Figure 8.9 *Ripples*

The waves that travel away from the source carrying energy are called **progressive waves**. The surface of the water is the medium in which they travel.

Each water drop in the water surface moves up and down as the wave passes. Energy is passed on from drop to drop.

Place a small piece of cork to float on water in a tank. Send waves across the tank. The cork moves up and down; that is across the direction in which the wave is moving. The waves which have particles moving *across* the direction in which the wave is travelling are called **transverse waves**. Water waves are transverse waves.

If you push and pull one end of a slinky spring, a wave is sent along the spring. The parts move backwards and forwards, in line with the direction in which the wave is moving. These are **longitudinal waves**.

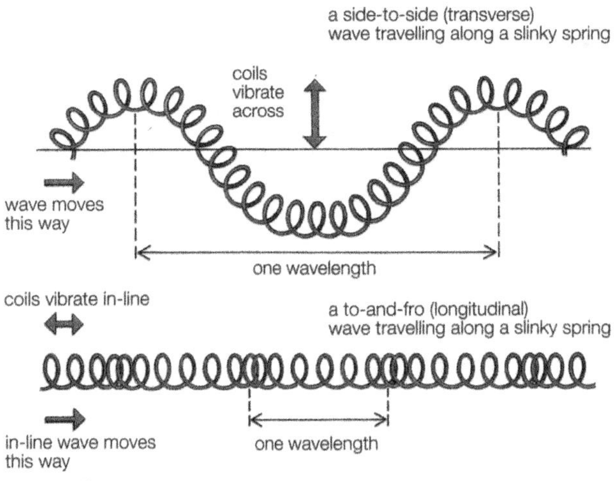

a side-to-side (transverse)
wave travelling along a slinky spring

coils
vibrate
across

wave moves
this way

one wavelength

coils vibrate in-line

a to-and-fro (longitudinal)
wave travelling along a slinky spring

in-line wave moves
this way

one wavelength

Figure 8.10

When waves move, the particles of the medium simply vibrate about an average position. They pass their motion and energy on to their next neighbour. The wave has moved on but the molecules stay where they were.

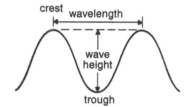

Figure 8.11

A wave has a high point called a **crest** and a low point called a **trough**. The height from the trough to the crest is called the **wave height**. The **wavelength** is the distance between two crests. Some waves can have a wave height of 30 m. In the ocean the wavelength of a wave is usually about 20 to 30 times its height.

As a wave approaches the shore, the water from far out in the sea is not moving in. As a wave crest reaches a particle of water, the particle moves upwards and forwards very slightly. As the wave crest passes, it moves downwards and backwards very slightly. The water particles, therefore, remain in about the same place. It is only the rolling motion, the wave shapes, that move in towards the shore.

As a wave comes towards the shore, its trough begins to touch the bottom, and this slows down the wave. As the trough touches the bottom, the wave crest

Figure 8.12 *Crashing waves*

becomes steeper. When the wave becomes too steep to support itself, the crest folds over and breaks into bubbly surf. As it folds over, the crest is called a **breaker**.

After the wave breaks, the water rushes back and runs under the next wave which is rolling forward. As it rushes back, it pulls sand along with it. The backward pull is called an **undertow**. Sometimes this undertow can be very strong and can pull out or even drown the best swimmers.

When you feel strong undertows, leave the water at once!

Earthquakes on the sea bed produce very long, low waves called seismic sea waves or tsunamis. Typical tsunamis have wavelengths of 150 km and travel at speeds of 12.5 km per minute. They can be very destructive when they hit the land.

Currents

Any force or additional movement in the water is called the **current**. The undertow caused by waves is an example of a current. Waves produce other kinds of currents.

When a large amount of water is piled up along the shore by waves, a **rip current** is formed. A rip current is a strong, narrow surface current that moves swiftly out from the shore for a short period of time. Rip currents can be dangerous to swimmers. If you are ever caught in a rip current, you should swim parallel to the shore until you are out of it.

Most waves strike shores at an angle, producing currents with a movement of water parallel to the shore. These currents are called **longshore currents**. They cause you to drift away from the place where you originally started to bathe. These currents also carry sand along the beach.

The effects of waves

Waves have a lot of energy and can do a lot of damage when they strike the shore. Waves cause **erosion** (wearing away of land) or **deposition** (building up of land).

When they strike land, waves cut into the shore forming sea cliffs and sea caves. Sea cliffs may eventually collapse into the sea. Waves also wash away the shore line.

Figure 8.13 *These rocks have been eroded by the sea*

Wave erosion produces a lot of sediment. This sediment, along with that brought by rivers that flow into the sea, is distributed by longshore currents and the undertow.

Longshore currents may deposit the sediments along the beach. The undertow deposits the sand offshore (away from the shore). These deposits may form **sand bars**.

Figure 8.14 *The Palisadoes across Kingston Harbour were formed by deposits from Hope River and elsewhere on Jamaica's south-east coast*

Ask your teacher to organize a field trip to the sea where you can carry out the following investigations.

8.2 The nature of the sea

Anyone who has bathed in the sea would have made at least one observation. The water is salty. The sea is the world's biggest salt solution. Scientists believe that the sea was first fresh water that became salty. As the rivers on land flowed over the soil and rocks, mineral salts were dissolved in them. These mineral salts were brought to the sea.

The water in the sea also evaporates and makes the salt solution even more concentrated.

The concentration of salt in the sea is called its **salinity** or **salt content**. The salinity of the sea varies from place to place, but the average salinity is 3.5 per cent. This means that in a 100 g sample of sea water, there is 96.5 g of water and 3.5 g of salt. The salt present is mainly common salt, or sodium chloride. Other salts present in small amounts include magnesium, potassium and calcium chlorides and sodium sulphate.

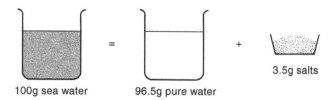

100g sea water = 96.5g pure water + 3.5g salts

Figure 8.15

Sea water also has oxygen and carbon dioxide dissolved in it. The oxygen is used by sea-dwelling plants and animals for respiration, while the carbon dioxide is used by the chlorophyll-possessing sea plants for photosynthesis.

Sea water is transparent and therefore allows light to penetrate it. How far down the light goes depends on how clear the water is, and on the sky and sea conditions. However, the greatest depth to which light can penetrate is about 150 m. The part of the sea to which light reaches is called the **light zone**. Below the light zone, there is darkness. Light is needed for photosynthesis, so plants which photosynthesize cannot live below the light zone. In the Caribbean most photosynthesis takes place in the top 10 m of the sea.

The temperature of sea water also varies. The temperature of the sea depends on latitude and depth. The surface temperature of the sea at the equator is higher than that at the polar regions. The temperature of the sea in deep water is close to freezing point, both at the equator and near the poles.

Since the sun's rays do not penetrate very far into the water, deep water is not warmed by the sun.

The importance of the sea

The sea contains large numbers of plants and animals which establish a food chain. The very small plants called **phytoplankton** carry out photosynthesis. Animals living in the sea eat the phytoplankton and are in turn eaten by other animals. For example, some phytoplankton may be eaten by an immature crab or a shellfish; this may be eaten by a small fish which in turn may be eaten by a larger fish which may be eaten by an even larger fish or a human.

On your own

1. Make a list of the food you eat that comes from the sea.

2. Compare the nutrients present in fish with those present in chicken or beef. (This will require you to do some research in the school library.)

3. Try to get, from the local fisheries office, information on the amount of fish caught in your territory in a given period of time.

4. Interview your local fishermen. Find out the type of fishing methods they use, the most common types of fish they catch, and the types most preferred by the public.

5. Find out at what time of year fish are most abundant and what the best time of the day for fishing is.

Figure 8.16 *Fishing using nets in Grenada (top) and the fish for sale (bottom)*

Figure 8.17 *This marine drilling platform is in the Gulf of Paria between Trinidad and Venezuela*

Oil contains carbon and hydrogen – elements found in all living things. It is thought that oil comes from tiny plants and animals which died long, long ago. Their bodies sank to the sea bed, and were covered with sand and mud which later hardened into rocks. Gradually, the action of chemicals in the rocks changed the bodies of the dead organisms into oil.

Humans have eaten seafood for centuries. Fish, shellfish and other animals from the sea have been an important source of food. They are a good source of protein, minerals and oils. However, humans must be careful not to overfish; that is, take out too many fish at one time or too often. If they do, the fish may not have time to reproduce, and their numbers will fall.

The sea also contains large amounts of mineral deposits and petroleum. Large quantities of manganese, copper, cobalt and nickel are deposited on the ocean floor. As supplies on land are used up, humans may turn to the sea for these minerals.

Large reservoirs of oil and natural gas are found in the floor of the sea. Special drilling platforms have been designed to get the oil and natural gas from the sea floor (see Figure 8.17). Trinidad and Tobago gets most of its revenue from oil and natural gas extracted from the sea bed.

Some experiments have been successful in producing electricity from both wave action and heat energy in the sea.

In some parts of the world, where there is a shortage of fresh water, the salt is removed from the sea water and the water used for irrigation and drinking. The removal of salt from sea water is called **desalination**. There are desalination plants in Antigua and the Netherlands Antilles, see Figure 8.18.

Figure 8.18 *A desalination plant in Curaçao*

159

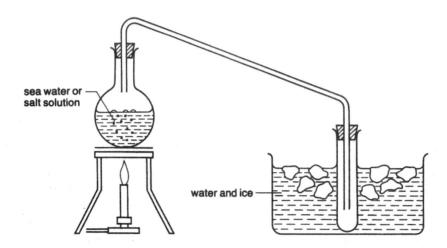

sea water or salt solution

water and ice

Figure 8.18

ACTIVITY 8.2 DISTILLING SEA WATER

You are going to separate sea water into pure water and salt by distillation.

You will need the apparatus shown in the diagram above.

1. Set up the apparatus as shown in the diagram. Use the sample of sea water you collected on your trip to the sea. If you do not have any sea water, make a strong salt solution and use it instead.

2. Heat the flask until you have collected half a test tube of water.

   What does the water you produce taste like?

You have separated water from the salt solution using the method of **distillation**. When the solution is heated, water evaporates, but the salt does not. The water vapour travels along the tube to the test tube, where it is cooled and condenses again. The condensed water contains no salt.

What disadvantage do you think there is to obtaining fresh water in this way on a large scale?

In the Caribbean, the sea is an important means of recreation. Many tourists come to enjoy the sea and our lovely beaches. Many of us go to the sea regularly: not as often as we would like, but whenever we can. Spending a day by the sea and bathing is a very relaxing and healthy pastime. We get exercise by swimming and we can relax and rest on the beach.

Avoid too much direct sunlight on the beach because the rays of the Sun can damage the skin and cause skin cancer.

8.4 Polluting the sea

As well as taking food out of the sea, unfortunately humans dump unwanted substances – **pollutants** – into the sea. As a result of our lifestyle we produce a lot of waste. Waste is produced by industry, shipping, irrigation, recreation and household activities such as washing and, with ever-increasing populations, a lot of sewage needs to be disposed of. Large quantities of this waste have been dumped in the sea and it is affecting the organisms living there. This in turn may affect the number of tourists who visit areas such as the Caribbean which has a rich marine environment.

Figure 8.19 *Relaxing by the beach*

Pesticides, e.g. DDT which is used to kill pests on land, find their way into the sea via rivers. High levels of these pesticides have been found in sea birds, which get them from the fish they eat. Fertilizers used on farm land also reach the sea and raise the nitrate levels.

Oil spills also cause a great deal of harm to organisms in the sea. Crude oil and water do not mix. As the oil floats on the sea, it blocks out the light needed by the plants to make food. The plants, and the animals that depend on them for food, will die. When the plants die, less oxygen is available for the animals to breathe. Sea birds become covered with oil, and they are unable to fly so they starve or drown. The food chain is destroyed.

Figure 8.20 *An oil spill that has come ashore and polluted the beach*

In order to preserve the sea, we must find new methods of waste disposal. We must try to use materials that decompose (break down) and can therefore be buried and recycled. Plastics take a very long time to break down.

We must develop pesticides that break down quickly into harmless substances, so that they do not remain and pollute the sea.

Industries must find ways of converting their waste into a form that is not harmful to organisms in the sea and we must treat our sewage.

Ways must be found of preventing oil leaks from drilling platforms and oil spills from tankers. We must develop efficient and effective methods of cleaning up oil spills when they occur. Tankers must be stopped from cleaning their tanks at sea.

The sea is important to us, let us preserve it.

Summary

Here are some of the important ideas you learnt in this unit:

- Tides are caused by the force of gravity on the Moon and the Sun.

- The most important influence on the tides is the Moon.

- The inter-tidal zone is a special habitat to which some plants and animals are adapted.

- Waves represent a transfer of energy.

- In transverse waves the particles move across the direction in which the wave is travelling.

- In longitudinal waves the particles move in line with the direction in which the wave is moving.

- Waves are produced when the sea is disturbed by the wind.

- The action of waves alters the appearance of the coastline by erosion and deposition.

- The sea has salts, oxygen and carbon dioxide dissolved in it.

- Photosynthesizing plants can live in the light zone in the sea.

- The sea is an important source of food for humans.

- We extract valuable minerals, oil and natural gas from the sea floor.

QUESTIONS

1 The difference in the height between the water level at low and high tide is called

   A   ebb and flow
   B   inter-tidal zone
   C   tidal range
   D   flood level.

2 What effect would waves have on the movement of a small fishing boat floating far out in the sea?

   A   The boat would be carried nearer to the shore
   B   The boat would move up and down with the waves
   C   The boat would drift further out to sea
   D   The boat would capsize.

3 Neap tide occurs when the sun and moon are
   A   at right angles to each other
   B   in line with each other
   C   opposite to each other
   D   parallel to each other.

4 What provides the energy to form waves in the sea?
   A   Earth movement
   B   The movement of ships
   C   The movement of the Sun
   D   The wind.

5 Which of the following represents a marine food chain?
   A   shellfish ⟶ smallfish ⟶ phytoplankton ⟶ large fish
   B   small fish ⟶ shellfish ⟶ phytoplankton ⟶ large fish
   C   phytoplankton ⟶ shellfish ⟶ smallfish ⟶ large fish
   D   phytoplankton ⟶ small fish ⟶ shellfish ⟶ large fish

6 What is a spring tide? Explain how spring tides occur. Use diagrams to illustrate your answer.

7 Why is sea water salty? Explain the presence of salt in sea water.

8 Explain the presence of fossil fuels in the sea bed.

9 Explain the ways in which the sea is important to people in the Caribbean.

10 On Earth we use large amounts of water in the home, in industry and in other areas, yet water is still present on the planet. Explain all the processes involved that ensure we have a supply of water on Earth. Use illustrations with your answer.

Natural resources

OBJECTIVES

- Understand what a natural resource is
- Learn about classification of natural resources
- Understand the need to conserve our natural resources
- Identify the benefit and burdens of the use of our natural resources

Every day we use products made from Earth's natural resources. What are these natural resources and how do we classify them? What are the benefits and burdens we face with the use of the natural resources?

Figure 9.1   *Sources of alternative energy*

9.1 Natural resources

Did you know that those living and non-living things that we obtain from the environment are called natural resources? Examples of some of these natural resources are water, air, living space and the Sun. Natural resources are classified as **renewable** and **non-renewable**.

Renewable resources

When these are used up, they are replaced by natural processes that occur over short periods of time such as months, years or decades. Examples of renewable resources are plants and animals used for food, natural fibres for clothing, and trees for lumber or paper. Energy from the sun, wind and flowing water is also considered to be renewable.

Figure 9.2 *In Costa Rica, banana stalks are now being used to manufacture paper, instead of disposing tons of the stalks into rivers and land fills where they would be harmful to the environment*

Non-renewable resources

These are formed in the Earth's crust but the processes that produce them take a very long time. Significant deposits of these resources take millions of years to accumulate, so we consider Earth to have a limited supply of them. This means that once they are used up, there will be no more available. These resources are described as **non-renewable**. Fossil fuels – coal, petroleum, natural gas, and important metals such as aluminium, iron, gold and copper are non-renewable resources.

Some non-renewable resources can be used more than once, as in the case of aluminium cans. They are crushed and reused in other aluminium products, see Figure 9.3. This type of reuse is called **recycling**.

Renewable or non-renewable

Sometimes a resource is described as being renewable or non-renewable depending on its availability. One such resource is groundwater. When its rate of removal from underground matches the rate at which it is replenished, then that groundwater is referred to as a

Figure 9.3a *Aluminium cans*

Figure 9.3b *Crushed aluminium cans for recycling*

Figure 9.3c *Land that was once a bauxite mine. The soil has been replaced and trees have been planted to rehabilitate the land*

renewable resource. However, when it is removed in greater amounts than it is replaced, causing the water table to drop considerably, it is considered to be a non-renewable resource.

Human beings, more than any other species use the natural resources of Earth. This fact has serious implications as Earth's human population continues to rise. At the beginning of the nineteenth century the human population was one billion. It is predicted that by the year 2005 there will be seven billion people here on Earth.

What are the implications for this population growth? The answer is an increased demand for more resources. Along with the growth in population, there is also the need to have improved standards of living. The questions scientists are asking are 'how long will Earth with its limited resources be able to support its human inhabitants? What will happen when the present supplies are used up?'

Other questions that scientists are asking are 'how much destruction of the environment are we willing to accept as we use more and more of Earth's resources?' and 'can we find alternatives to these resources?'

One answer is that we must manage our resources wisely, that is they must be **conserved**. Conservation is the wise use of renewable and non-renewable resources so that they will not be used up, or cause damage to our environment when we use them.

One method of conserving metal resources is recycling. In some Caribbean countries, public organizations such as the Caribbean Conservation Association inform the public about the three **R**'s of conservation: **R**euse, **R**educe and **R**ecycle.

Burning of fossil fuels

Human use of the Earth's natural resources has led not only to improved standards of living but also to serious environmental problems. One main example is the use of fossil fuels. This natural resource is used for transportation, in industries, by agriculture and by power stations. But when fossil fuels are burnt, a range of air pollutants are released into the atmosphere. Some of these pollutants are gases such as oxides of sulphur (SO_2 and SO_3), oxides of nitrogen (NO and NO_2) and solid airborne particles of carbon called soot.

The oxides of sulphur and nitrogen react with water vapour and make weak acids which fall to Earth as acid rain. Acid rain changes the pH of the natural habitats of plants and animals, causing damage, destruction and death to the animals and plants that live on the soil and in water.

Gases such as SO_2 also irritate the respiratory tracts and eyes of some people, and can also cause them to have asthmatic attacks or serious allergic reactions.

Acid rain reacts with buildings made of marble and limestone, and causes these materials to deteriorate over a period of time. Acid rain also causes metals to corrode. Buildings made of limestone and with metal fixtures are thus seriously affected by acid rain.

Figure 9.4 *Acid rain causes deterioration of buildings*

Carbon dioxide and global warming

Scientists have found that the level of carbon dioxide in the atmosphere has been increasing significantly, and they have linked this increase to the burning of fossil fuels.

Carbon dioxide is a heat absorber, so it absorbs much of the radiant heat reflected from Earth's surface, causing an increased temperature of the lower atmosphere and leading to global warming.

Changes in weather patterns have led to an increase in rainfall and floods in some countries of the world, and prolonged droughts in others. Scientists also attribute the El Niño effect, which is the warming of the waters of the eastern Pacific Ocean off Ecuador, Peru and northern Chile, to global warming.

9.2 Some resources

Land as a resource

The survival of human beings depends on our having enough land to grow food. Every year many thousands of people die from starvation. About 70 per cent of the world's population do not get enough to eat. Even with the great demand for food, every year valuable land all over the world is used to provide houses, factories, roads and airports. Fertile valleys are flooded to supply water and power. Careless farming causes soil erosion and thereby creates new deserts.

As populations increase, more and more land is used up. The habitats of wild plants and animals are destroyed. Many species are in danger of becoming extinct. How can we relieve this problem?

Conservation of land

One of the methods of conserving our land resources is to bring new land into food production. Irrigation schemes, where possible, can make some new land available for growing food.

Derelict land such as old quarries, mines and dumps can sometimes be reclaimed for building. This helps to take the pressure off agricultural land which might otherwise have been built on.

Laws can be passed to prevent building on farm land around towns and cities. These areas are called **green belts** and have been traditional areas of food supply. Other areas such as national parks, bird sanctuaries and nature reserves have been set aside to preserve natural habitats and the organisms which live there.

Figure 9.5   *The Asa Wright Centre in Trinidad is a nature reserve*

Mineral resources

Land also serves as a source of minerals which may yield metals or non-metals. While some metals such as gold and silver occur naturally in a chemically uncombined form, most metals and non-metals are found in a chemically combined form called **minerals**. Rocks in which deposits of minerals are found are called **ores**.

Bauxite is an example of an ore. From bauxite is extracted the metal aluminium. Guyana, Jamaica and Suriname have significant deposits of bauxite.

Soil as a resource

Soil forms the habitat for a large community of different organisms. This community is affected by the soil in which it lives and in turn the organisms affect the soil. The interactions between a habitat and its community make up an **ecosystem**.

Soil contains minerals. These minerals come from the rocks which break down to form the soil. Weak acids in the rain dissolve some mineral salts out of the parent rock and these end up in the soil water. As plants grow in the soil, they remove the dissolved minerals from it. If this process continues for a long time, the soil may become exhausted of its minerals so that, eventually, plants can no longer grow in the soil.

Minerals may also be washed out of the soil by rainwater. This is called **leaching**. Leaching is a problem in fast draining sandy soil. Given a chance, Nature will replace much of these lost minerals. Dead plants and animals and their wastes are all rich in minerals which came originally from the soil. When these remains mix with the soil, they are broken down by the **decomposers** to form humus. Gradually, as they decay completely, their minerals are returned to the soil water to be available once more to plants.

When a field or garden is used to grow crops, this kind of natural replacement of minerals does not happen. When the crop is harvested, the plants with the minerals they contain are removed from the ground for good. The lost minerals should be replaced by adding dead plant and animal material, e.g. farmyard manure, garden compost, leaf mould or bone meal, to the soil. Artificial fertilizers, e.g. ammonium sulphate, can also be used.

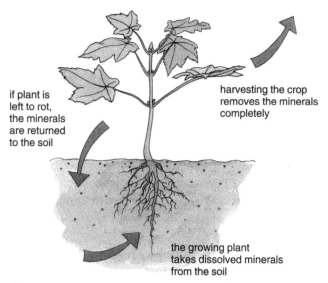

if plant is left to rot, the minerals are returned to the soil

harvesting the crop removes the minerals completely

the growing plant takes dissolved minerals from the soil

Figure 9.6

Crop rotation, which involves growing a different crop each year, should also be practised. Crop rotation ensures that plants which need a lot of one particular mineral do not exhaust this mineral too quickly from a field. Different crops, on the whole, extract a different range of minerals from the soil.

Soil erosion

Bad farming methods can lead to topsoil being carried off by wind or water. This is called **soil erosion**. Soil erosion leaves the land barren and useless.

Wind erosion is erosion of the soil caused by wind. When no organic fertilizers are used, the humus content in the soil gradually declines; the soil becomes dry and light and is easily blown away in dry weather. When hedges and trees are removed, there is nothing to break the force of the wind. The wind, therefore, takes away the topsoil. If poor and dry pastureland is overgrazed, wind erosion is likely.

If a slope is ploughed in an 'up or down' direction, channels are produced when rain falls and the soil is carried away. Deep gullies are carved in the soil which makes the land useless for farming. Gully erosion can be prevented by ploughing the land across, rather than up and down, the slope. Ploughing across the slope is called contour ploughing.

Trees growing on hillsides retain water and hold the soil in place. If trees are removed – **deforestation** – the soil is washed into the rivers. These become clogged and floods result. Replanting trees – **re-afforestation** – and contour ploughing reduce the loss of topsoil.

Water as a resource

Human health and survival depend on plentiful amounts of fresh, clean water. As populations increase, so does the demand for water. It is estimated that each person drinks about 1.5 litres of water per day. We also use water for other things, e.g. bathing, cooking and cleaning. Each person uses approximately 400 litres of water per day. In addition, industry uses large quantities of water.

In many parts of the world, where rainfall is low, poor water supplies make life very difficult. Even countries which have plenty of rain may have difficulty in meeting the demand for water. Increasing domestic and industrial use, inadequate collection of water, and pollution of water sources are some reasons for the inability of many countries to meet the demand for water.

Water conservation

Water is a remarkable resource. The water we throw away is cleaned and recycled for us by Nature. This is known as the **water cycle** – see Book 1, Unit 5.

Water conservation is very important. We must realise that water is not limitless. We must think of ways to conserve water. One obvious way is to make sure that taps (faucets) in our homes do not leak. In what other ways can we help conserve water? Make a list.

Water for drinking and bathing must be purified. **Water purification** is the process by which water is cleaned so that it is usable. Water purification takes place in special situations. The water is filtered to remove particles. Then it is chemically treated to kill bacteria and other dangerous organisms. Many cities and towns have treatment plants to purify drinking water. Water can be a source of pathogens (germs) that cause very serious diseases (see Book 1, Unit 7). Many large industries have treatment plants to clean waste water before it is released into lakes and streams. This prevents pollution of these water sources.

Sewage

Raw sewage consists of human faeces and all the other wastes which we flush down our toilets and sinks. It provides a perfect food for bacteria and other microbes which can quickly convert it to harmless substances. This is what takes place in our 'soakaway' at home and on a larger scale in sewage plants.

In sewage plants the bacteria are allowed to convert the sewage to harmless substances on gravel beds supplied with plenty of air. The treated sewage can then be pumped safely into a river or the sea. Very often sewage plants cannot cope with the large supplies and they are forced to discharge raw sewage.

Figure 9.7 *Sewage treatment plant*

As the raw sewage is pumped into the waterway, bacteria in the river multiply rapidly. The large population of bacteria absorb all the oxygen from the water. Soon many organisms living in the water die because they lack oxygen.

Sometimes a river can recover and continue to have a great variety of organisms. Unfortunately, however, the river in most cases becomes one large sewer.

A useful indicator that water has been polluted with human faeces is the presence in it of a bacterium called *E.coli*. This organism is always present in faeces. Water containing *E.coli* is unfit to drink.

9.3 Fossil fuels

Buried deep within the Earth's crust lie the sources of much of the energy we use daily. These natural resources are formed from the remains of living things. They have developed and accumulated over long periods of time earlier in the Earth's history. These natural resources are formed from the remains of living things and are called **fossil fuels**. Examples of fossil fuels are coal, petroleum and natural gas.

Fossil fuels consist mainly of compounds of carbon and hydrogen called **hydrocarbons**. These compounds contain energy originally obtained from sunlight by plants and animals that lived millions of years ago. Energy can be released from hydrocarbons, when these are burnt, in the form of heat and light.

Coal

Coal is made up almost entirely of plant material. The plant material goes through four stages in the formation of coal. In the first stage of coal formation, **peat** is formed. Peat is made of decaying leaves, twigs and branches. It has a high water and carbon content. As it burns, peat releases smoke and pollutants.

Heat and pressure change peat into **lignite**. This is more solid than peat. It is a soft brown coal with about 40 per cent carbon. When lignite burns, it releases pollutants that are harmful to people and the environment.

Over time, more pressure and heat convert lignite into **bituminous coal**. This soft coal is about 85 per cent carbon. It is the most common type of coal in the world. Because of its high carbon content it burns cleanly and releases fewer pollutants.

Great pressure and heat, over millions of years, change bituminous coal to **anthracite** or hard coal. This is 90 per cent carbon. When it burns it releases a

Figure 9.8   *Peat*

great amount of heat but few pollutants. Unfortunately anthracite is not very common.

Petroleum and natural gas

Petroleum and natural gas are hydrocarbons formed largely from microorganisms that lived in oceans or large lakes millions of years ago. Petroleum, also called oil, consists of liquid hydrocarbons. Natural gas is made up of hydrocarbons in gaseous form.

When microorganisms die in shallow prehistoric oceans and lakes, their remains accumulate on the ocean floor and lake bottoms and are buried by sediments. These sediments limit the available oxygen supply and prevent the remains from decomposing completely. As more and more sediments accumulate, the heat and pressure on the buried organisms increase. When the heat and pressure become great enough, chemical changes occur that convert the remains into petroleum and natural gas.

Petroleum and water are usually trapped between layers of impermeable rock – rock through which liquids cannot flow. Because petroleum is less dense than

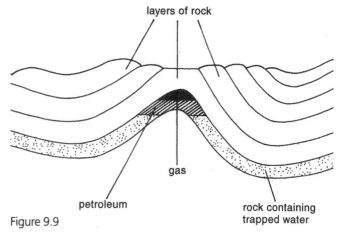

Figure 9.9

water, it floats on top of the trapped water. If natural gas is present it is found on top of the petroleum because natural gas is less dense than both oil and water. When a well is drilled into an oil pool, the petroleum and natural gas often flow to the surface.

Uses of fossil fuels

Fossil fuels are the main sources of energy for industries. At one time coal was the major source of energy throughout the world. It was used to heat buildings, drive locomotives and ocean liners and run factories. Much of the coal used today produces electricity. At power plants the heat from burning coal boils water to produce steam. The steam turns huge fan-like devices called turbines, in electricity generators. The generators change the mechanical energy of the turbines into electrical energy.

Petroleum and natural gas meet the energy needs of most countries. Petroleum pumped from a well is refined into various products. Many types of transportation use petroleum fuels. These include gasoline, diesel fuel, kerosene and jet fuel.

Many non-fuel products such as motor oil, waxes, asphalt and petrochemicals also come from petroleum. Petrochemicals are used to make plastics, synthetic fabrics, medicines, building materials, synthetic rubber, insecticides, chemical fertilizers, detergents and shampoos.

Fossil fuel supplies

Fossil fuels are non-renewable resources. Coal is the most abundant fossil fuel in the world. However, scientists estimate that at its present rate of use, the world-wide coal reserves will last only about 200 years.

The United States has been explored for petroleum more completely than any other country in the world. Scientists estimate that at least 75 per cent of all the petroleum in that country has already been discovered. In other areas of the world more than 90 per cent of the petroleum known to exist is still in the ground. Scientists also think that much natural gas remains to be discovered, but it lies more than 4600 m below the Earth's surface. We must conserve our fossil fuels!

Fossil fuels and the environment

The use of any fossil fuel has an impact on the environment. Strip mining leaves deep ditches where soil is removed. Rocks and topsoil that are displaced to expose the coal are left lying on steep slopes. Without plants and topsoil to protect it, the exposed level is then eroded.

When wet, rocks exposed during mining give off acids. Rain carries the acids into nearby rivers and streams causing harm to living things there.

Fossil fuels can also cause air pollution.

Spills from oil wells, tankers or pipelines pollute the ocean and harm plants and animals.

a

b

c

d

e

f

g

Figure 9.10   *Butane, tyres, paint, plastics and deodorants are products of petroleum*

9.4 Alternative energy sources

No matter how much conservation of fossil fuels takes place, sooner or later the Earth will run out of these non-renewable resources. Scientists are hard at work developing alternative energy sources. Many of these alternative energy sources are used to generate electricity.

Solar energy

Solar energy is energy from the Sun. It is non-polluting and almost unlimited.

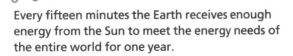

Every fifteen minutes the Earth receives enough energy from the Sun to meet the energy needs of the entire world for one year.

Large **solar collectors** located on the roofs of buildings can convert the sun's energy into heat to warm the building and provide hot water. **Solar cells** can also be used to transform light energy directly into electricity. However, these cells are expensive and produce only small amounts of electricity.

Geothermal energy

Geothermal energy is heat energy from within the Earth. Hot igneous rock near the Earth's surface heats underground water and changes it into steam. Geothermal reservoirs form when rock traps the hot water and steam underground. These sources can be drilled to release the hot water and steam. The steam is fed into electricity generating plants. Several plants around the world are now producing electricity from geothermal energy. In Iceland, almost 80 per cent of the homes get their heat and hot water directly from hot springs and geysers. Is geothermal energy renewable or non-renewable?

Energy from water

Water moving through huge dams built across rivers turns turbines and generators that produce electricity. This is **hydroelectric energy**. At the present time about one-quarter of the world's electricity is produced in this way. However, dams used for hydroelectric plants may cause environmental problems – particularly if they suddenly burst.

Figure 9.11   *A geothermal power station in New Zealand*

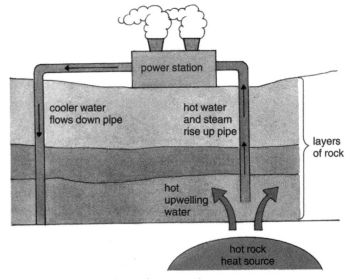

Figure 9.12   *A geothermal energy scheme*

In some coastal areas, the water of an incoming tide rushes into various rivers or bays. When the tide changes, the water rushes with great force back to the ocean. A hydroelectric plant can be built to use this movement of the water. **Tidal energy** is used in France and eastern Canada. It requires special conditions, and therefore its use may never be widespread.

Energy from wind

Small wind-driven generators are used to meet some energy needs in individual homes. Larger devices produce power to meet most electricity needs in certain locations. However, even in the most favourable locations, the wind does not always blow. Therefore wind-driven generators work best if some electricity is stored. Snow and freezing rain can interfere with the operation of windmills.

Nuclear energy

Vast amounts of energy can be released when the nucleus of an atom is split. In a nuclear power plant, atoms of uranium are split in a nuclear reactor. The process is called **nuclear fission**. When the nucleus is split, the heat given off can be used to convert water to steam. The steam is then sent through turbines to produce electrical energy.

While producing large amounts of energy, nuclear fission also produces radioactive waste which is extremely dangerous. To date, no safe way of storing nuclear waste has been discovered. The waste is being stockpiled.

Another danger of nuclear power is that, if there is an accident at the plant, radiation may be released. In addition, the costs of building nuclear plants have risen tremendously. Because of the costs and the danger of radiation, very few new nuclear power plants are planned.

Scientists will continue to investigate ways of finding alternative sources of energy as we attempt to conserve the fossil fuels.

ACTIVITY **9.1** **CALCULATING THE COST OF ELECTRICITY**

You are going to find out the cost of electricity and suggest ways of reducing the amount of electricity used.

1 On a Monday morning read the electricity meter in your house before going to school. The next day, at the same time, read the meter again. Determine the number of kilowatt-hours used in that 24-hour period.

2 Select four other 24-hour periods, including weekends, and determine the number of kilowatt-hours of electricity used in each.

3 Calculate the average number of kilowatt-hours of electricity used in one day.

4 Call your local power company to find out the cost per kilowatt-hour. Calculate the cost of electricity for one day, one week, one year.

5 Suggest ways in which you can reduce the amount of electricity used in your household.

QUESTIONS

1 World demand for energy is
   A decreasing
   B increasing
   C constant
   D over.

2 Geothermal energy comes from heat stored in hot
   A water
   B wind
   C rocks
   D weather.

3 Coal is formed over millions of years from
   A the remains of organisms
   B bacterial activity
   C fossil fuels
   D mineral deposits.

4 The removal of the topsoil is called
   A reclamation
   B landfill
   C soil erosion
   D leaching.

5 Aluminium can be taken out of bauxite, which is
   A an ore
   B an energy source
   C a renewable resource
   D a fossil fuel.

6 Energy resources that have formed from the remains of living things are called
   A minerals
   B metals
   C renewable resources
   D fossil fuel.

7 The splitting of the nucleus of an atom to produce energy is called
   A geothermal energy
   B nuclear fission
   C nuclear fusion
   D hydroelectric power.

8 Is tidal power renewable or non-renewable? Explain.

9 Name and describe the four stages of coal formation in the order in which they occur.

10 How would your life be different without fossil fuels? Explain.

End of term test 1

Units 1, 2, 3

Select the letter which best answers the questions:

1 An element common to all acids is
   A chlorine
   B hydrogen
   C nitrogen
   D oxygen.

2 When salt dissolves in water, the water is the
   A solute
   B solvent
   C solution
   D substance.

   For questions 3 and 4
   A Chromium
   B Carbon
   C Helium
   D Carbon dioxide

3 Which of the substances listed above is a compound?

4 Which of these is both an element and a gas?

5 Which process can be used to obtain water from sea water?
   A Chromatography
   B Distillation
   C Evaporation
   D Filtration

6 Work is defined as the product of
   A force and distance moved
   B force and weight
   C force and energy
   D force and surface area.

7 Which of the following simple machines is a wheelbarrow an example of?
   A Lever
   B Pulley
   C Wedge
   D Wheel and axle

8 Which of the following is characteristic of liquids? The molecules
   A are arranged in a definite order.
   B are of one type only.
   C slide past each other.
   D are large in size.

9 Which of these statements is the best definition of a compound?
   A Can be a solid, liquid or gas combined.
   B Is a pure substance.
   C Consists of more than one type of atom chemically combined.
   D Is the building block of natural substances.

10 Which of the following is *not* a force?
   A Friction
   B Gravity
   C Mass
   D Weight

11 a Explain the meaning of these terms:
      (i) solute
      (ii) solvent
      (iii) solution
      (iv) saturated solution
      (v) dilute.

   b Name a well-known solvent which can
      (i) remove tar from your shoe,
      (ii) remove red dye from your hand.

c Describe in detail an experiment you would perform to compare two products which claimed to remove grass stains from white cricket trousers.

12 a The diagram below shows which elements are found in the Earth's crust (the outer shell of the Earth). The figures show what percentage of the Earth's crust each element accounts for. Study the diagram and then answer the questions below.

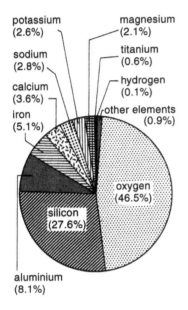

(i) Write the symbol for each element in the diagram.
(ii) Which of the elements are metals?
(iii) Which of the elements are non-metals?
(iv) Which of the elements are gases?
(v) Which is the most abundant (most common) element in the Earth's crust?
(vi) Which is the most abundant metal in the Earth's crust?
(vii) In which substance is silicon found?
(viii) Oxygen is found in almost every substance on Earth. Why do you think this is so?

b (i) Imagine that you could cut up a lump of iron sulphide into the smallest piece that shows the properties of iron sulphide. What atom or atoms would that piece of iron sulphide contain? Is iron sulphide an element or a compound?

(ii) Imagine you could cut up a lump of copper into the smallest piece that shows the properties of copper. What atom or atoms would that piece of copper contain? Is copper an element or a compound?

(iii) Draw two columns in your note book. Name one column 'compound'; name the other 'mixture'. Place each of these substances in the correct column: air, water, brass, salt water, sodium chloride (cooking salt), sugar, sea water.

13 A spring has a length of 3.6 cm when no weight hangs from it. When a weight of 50 N hangs from it, its length becomes 8.6 cm. The graph below shows the relationship between the two variables.

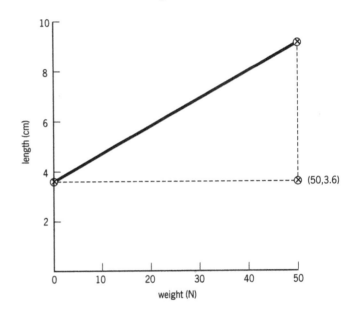

a Write down the mathematical equation that can be used to represent the relationship between the two variables.

b (i) Write down the co-ordinates of the y intercept.
(ii) Calculate the slope of the line.

c Use the equation and find the length of the spring when the weight is 20 N.

14 The diagram below shows an outline of the periodic table with 11 elements in position.

																S
											M					
F													D		T	
		R												B		
	Q			X												
						Z								L		

a Use the letters given for each element to answer the following questions.
Example: Give one metal element. Answer: R.

  (i) Give two elements in the same group.
  (ii) Give two elements in the same period.
  (iii) Which element is in group 1 of the periodic table?
  (iv) Which element is in group 7?

b Using a copy of the periodic table, find the name and symbol of each of the lettered elements given in outline. Show the letter, the name and the symbol in a table in your note book.

15 a Using what you know of the structure of the atom, fill in the gaps in the following table.

	Atomic number	Mass number	Number of protons	Number of neutrons	Number of electrons
Element A	30	65			
Element B		75		42	
Element C				18	17
Element D			19	20	
Element E		1			1

b Using your completed table and a periodic table, find the names of the elements in your table.
Below are diagrams of molecules. Give the formula of each.

(a)

(b)

(c)

(d)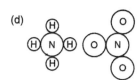

End of term test 2

Units 4, 5, 6

1 The instrument used to measure electric current is called:

A a resistor
B a rheostat
C an ammeter
D an electromagnet.

2 Fuses and circuit breakers have the same function. They:

A protect circuits when the current is too high
B protect circuits when the current is too low
C melt and prevent a fire
D break and allow the current to flow.

3 Ohm's law, stated as an equation, is:

A $I = VR$
B $V = IR$
C $R = IV$
D $C = VP$

4 In an electromagnet, when the current stops:

A it loses its magnetism
B its magnetism increases
C its magnetism decreases
D it becomes a neutral magnet.

5 A, B, C and D are four lengths of wire. A and C are the same length but C is twice as thick as A. B and D are the same length but D is twice as thick as B. A and C are equal in length while B and D are equal in length and twice as long as A and C.
Which length of wire will have the greatest resistance to electricity?
A B C D

6 Which of the following is *not* an electromagnet?

A Telephone
B Industrial crane
C Doorbell
D Compass

7 There are three ways in which heat energy can be transferred. Next to each of the following statements write the type of heat transfer it describes:

(i) Standing in front of a camp fire, you feel the heat _____
(ii) During the day the breeze usually blows from the sea to the shore _____
(iii) A metal rod with one end in a fire becomes hot at the other end _____

8 Fill in the blanks:

Substances are made up of _____ which are always moving. Since they are moving they have _____ energy. _____ moving particles have more energy than particles which are moving _____.
In solids the particles are very _____ packed and move backwards and forwards around the same position. In liquids the particles are _____ but slide passed one another. Particles of gases are far _____ and move _____ rapidly in _____ directions.

9 What is the difference between a solar eclipse and a lunar eclipse? Illustrate your answer with diagrams.

10 Explain the following:

(i) a swimming pool looks shallower than it really is when you look into it
(ii) a rainbow always has seven colours
(iii) the driver of a car can see behind him by looking in a mirror
(iv) some people have to wear spectacles (glasses) in order to see properly.

11 What are the uses of the following?

    A Microscope
    B Prism
    C Convex lens
    D Camera
    E Binoculars

12 What is an echo?
Describe one way in which echoes are used.

13 Identify the following parts of the human ear:

   (i) three little bones are found here
  (ii) takes impulses from the ear to the brain
 (iii) collects sound waves and channels them down
       the ear passage
 (iv) contains sensory hairs which are stimulated by
       vibrations
  (v) helps us to maintain our balance

14 What is the difference between short sight and long sight? How can these defects be corrected? Illustrate your answer with diagrams.

15 What is the specific heat capacity of a substance?
Below is a list of the specific heat capacities of a number of substances:

silver – 240 J kg^{-1} °C^{-1}
lead – 130 J kg^{-1} °C^{-1}
iron – 100 J kg^{-1} °C^{-1}
glass – 650 J kg^{-1} °C^{-1}

Plot these values on a graph.
Which one of the above mentioned substances would you use for a container to boil water? Give reasons for your answer.
If you were in the business of heating up substances and you had to charge by the amount of heat used, put in order, with the highest charge first, how you would charge for heating the four substances mentioned above. Give reasons for your answer.

End of term test 3

Units 7, 8, 9

1 Ploughing the land across the face of a slope is called

  A terracing
  B contour farming
  C crop rotation
  D strip ploughing.

2 Energy from the sun is called

  A geothermal
  B solar
  C nuclear
  D non-renewable.

3 Which member of a food web returns nutrients to the environment?

  A Producer
  B Consumer
  C Predator
  D Decomposer

4 Many animals in the Arctic have fur that is brown during spring, summer and fall but turns white in winter. This colour change is one way in which these animals are

  A protected from diseases
  B adapted to their environment
  C able to attract mates
  D able to conserve energy.

5 Which of the following is *not* a fossil fuel?

  A Coal
  B Natural gas
  C Sun
  D Petroleum

6 Which energy source is a renewable resource?

  A Coal
  B Natural gas
  C Moving water
  D Petroleum

7 The process by which salt is removed from ocean water is called

  A purification
  B detoxification
  C condensation
  D desalination.

8 Two major problems associated with the use of nuclear energy are thermal pollution and

  A environmental damage from strip mining
  B disposal of radioactive wastes
  C acid rain caused by air pollution
  D lung diseases in the population.

9 Global warming caused by air pollution results from

  A the El Niño effect
  B an increase in the world's population
  C the benefit and burden of technology
  D burning of increased amounts of fossil fuels.

10 Which of the following structures are found in plant cells but *not* in animal cells?

  A Cell wall and chloroplast
  B Centrioles and mitochondria
  C Cytoplasm and cell membrane
  D Vacuole and nucleus

11 Read the following carefully:

From AD 650 to AD 1650, the Earth's population doubled from 250 million to 500 million. This doubling of the Earth's population took 1000 years. Now the Earth's population doubles every 33 years. The population of the Earth in 1986 was about 4.5 billion. Using these figures, determine the population of the Earth in the years AD 2019, AD 2052 and AD 2085.

Present your results on a graph.

How many times greater will the Earth's population be in 100 years' time?

How would this increase in population affect our use for renewable and non-renewable resources?

12 In the Caribbean area, 75 per cent of the natural income of Trinidad and Tobago comes from oil and natural gas.

What do you think would happen if the oil and gas reserves were depleted?

Consider alternative measures that could be taken to extend the usability and life of current oil and gas wells.

13 Wood is a popular resource for fuel in many areas of the world. In some places, however, wood is no longer plentiful and people have had to meet their energy needs in other ways. Explain how a renewable resource like this can become a non-renewable resource.

14 Consider the town or village in which you live.

List areas of pollution in your town or village. Classify the areas of pollution under the following headings — air pollution; land pollution; water pollution and noise pollution.

Give your neighbourhood a rating between 0 and 10 for each area of pollution. Explain the rating you have given.

A rating of 10 would mean no evidence of pollution, whereas a rating of 0 would mean extreme amounts of pollution.

Suggest ways in which each type of pollution could be reduced in your town or village.

15 The following living things can be found in a forest ecosystem:

ground plants, caterpillar, lizard, iguana, spider, woodpecker, owl, mahogany tree, mouse, snake, monkey, hawk, snail, small bird, insect, aphid and decomposers (worm, beetle, bacteria).

(i) From the list above identify a producer, a herbivore, an omnivore, a carnivore, and a tertiary consumer.

(ii) Arrange the organisms into a branched and a numbered key.

(iii) Arrange them into as many food chains as possible.

(iv) Group them into a food web.

(v) Which organisms would have the largest numbers in its population? Which organism would have the least numbers in its population?

Answers

Unit 1 page 14

1 D   2 C   3 B   4 A   5 B
6 A

Unit 2 page 55

1 D   2 C   3 C   4 A
5 (i) Zn   (ii) O   (iii) Br   (iv) Na
   (v) K   (vi) Ag   (vii) Sn   (viii) N
   (ix) He
6 (i) Chromium   (ii) Cobalt   (iii) Copper
   (iv) Carbon   (v) Chlorine   (vi) Calcium

10(a)

10(b)

Unit 3 page 72

1 D   2 D   3 B   4 C   5 D   6 B

Unit 4 page 88

1 B   2 C   3 A   4 C   5 D

Unit 5 page 104

1 C   2 D   3 C   4 A   5 C

Unit 6 page 129

1 C   2 D   3 B   4 C   5 B   6 A

Unit 7 page 150

1 A   2 D

Unit 8 page 162

1 C   2 B   3 A   4 D   5 C

Unit 9 page 171

1 B   2 C   3 A   4 C   5 A   6 D
7 B

End of term test 1, Units 1, 2, 3 page 172

1 B   2 B   3 D   4 C   5 B   6 A
7 A   8 C   9 C   10 C

**End of term test 2, Units 4, 5, 6
page 175**

1 C   2 A   3 B   4 A   5 B   6 D

**End of term test 2, Units 7, 8, 9
page 177**

1 A   2 B   3 D   4 B   5 C   6 C
7 D   8 B   9 D   10 A

Index